New Paradigms in Healthcare

In the first two decades of this new millennium, the self-sufficiency of Evidence-Based Medicine (EBM) have begun to be questioned. The narrative version gradually assumed increasing importance as the need emerged to shift to more biologically, psychologically, socially, and existentially focused models. The terrible experience of the COVID pandemic truly revealed that EBM alone, while being a wonderful scientific philosophy and containing the physician's paternalistic approach, has its limitations: it often ignores both the patient's and physician's perspectives as persons, as human beings; it pays relentless attention to biological markers and not to the more personal, psychological, social, and anthropological ones, removing the emotions, thoughts, and desires of life, focusing on just the “measurable quality" of it.

Health Humanities, Medical Humanities and Narrative Medicine are arts intertwined with sciences that allow to broaden the mindset and approach of healthcare professionals. helping them to produce better care and more well-being.

Aim of this series is to collect “person-centered” contributions, as only a multidisciplinary and collaborative team can meet the challenge of combining the multiple aspects of human health, as well as the health of our planet, and of all the creatures that live on it, in a common effort to stop or reverse the enormous damage committed by humans during our anthropocentric era: a new paradigm of healthcare, education and learning to create a sustainable health system.

David Cerdio-Domínguez
María Elizabeth de los Ríos-Uriarte
Editors

Current Bioethics

Trends and Debates

Springer

Editors
David Cerdio-Domínguez
Anahuac Center for Strategic Development in Bioethics
Anahuac University Mexico
Mexico City, Mexico

María Elizabeth de los Ríos-Uriarte
School of Law
Universidad Autónoma de Querétaro
Querétaro, Mexico

ISSN 2731-3247 ISSN 2731-3255 (electronic)
New Paradigms in Healthcare
ISBN 978-3-032-25293-7 ISBN 978-3-032-25294-4 (eBook)
https://doi.org/10.1007/978-3-032-25294-4

This Springer imprint is published by the registered company Springer Nature Switzerland AG
The registered company address is: Gewerbestrasse 11, 6330 Cham, Switzerland

Foreword: The Bioethics Mentality—Origins and Trends

The term bioethics became a standard word by the end of the twentieth century due to its introduction by Van Rensselaer Potter and André Hellegers, although the theologian Fritz Jahr had used it some decades earlier.

It is useful to consider the bioethics discourse from three points of view: first, as a social movement articulated around inequalities in access to healthcare and a reflection on the part of the academic community about the misuses of science and technology; second, as a form of decision-making in matters concerning the impact of science and technology on human life; and third, as an academic enterprise with bioethics as a field of expertise that can be learned and taught.

The essential aspect of the bioethical mentality is the use of dialogue for formulating and solving moral dilemmas concerning the environment, human and animal welfare, and just distribution of resources. It complements the two main streams of ethical thinking in Western philosophy: deontology, a theory of duties, and teleology, a reflection on the consequences of human action. The philosophical approach derives from wider systems of thought proposed by prominent thinkers. The dialogical approach involves the creation of committees or commissions reflecting the need to take into consideration different beliefs and cultural differences.

Currently, bioethics encompasses a wide spectrum of disciplines and points of view. It addresses problems related to the natural environment, human health, social justice, human rights, and scientific practice. It has been instrumental in calling attention to issues not hitherto considered by a global audience and has impacted on "soft law" documents reflecting the need to respect life, to emphasize human dignity and autonomy, and to anticipate challenges posed by technological advances.

The current state of the use of the bioethical discourse reveals its pervasiveness and usefulness. However, dangers are also present. Confusion between legal and ethical analysis is sometimes present. Political abuse, spurious interests, lack of intellectual rigor, and other shortcomings are also evident.

The future of bioethical thinking should bring about a better perception and analysis of challenges, particularly after the emergence of powerful technological tools. It may also produce a better understanding of the role of science, technology, and politics in human affairs. Education at university or college levels will improve.

This collection of articles is a timely and profound contribution to bioethical discussions. It gathers insights from different cultural settings and analyzes different fields relevant to the construction of a bioethical mentality, far beyond a change in research and practice paradigms.

Fernando Lolas Stepke

Professor and Researcher
University of Chile and Central University of Chile.
Former Director of the Bioethics Program
of the Pan American Health Organization
and Member of UNESCO International
Bioethics Committee. Editor-in-Chief of Acta Bioethica.
International Distinguished Fellow
of the American Psychiatric Association,
Santiago, Chile

Contents

Part I Fundamental Bioethics

1 **Gender Perspective in Bioethics: Toward an Integral Reflection** 3
Jennifer Hincapie Sanchez and Fernanda de Blas López

2 **The Right to Know: Fertility Awareness and Bioethical Decision-Making** 17
José Alberto Castilla Barajas, María Fernanda Delgado Domínguez, Fernanda Pérez Rea Vela, Claudia Lerma, Abel Lerma, José Marcos Félix Castro, and Ana Lucila Villa Viruette

3 **CRISPR/Cas9 and the Legitimacy of Nontherapeutic Genetic Intervention** 33
José Tomás Alvarado, Paulo López-Soto, Alex J. Wolf, and Manuel J. Santos

Part II Clinical Bioethics

4 **Current Structures, Practice, and Impact of Clinical Ethics Consultation in Healthcare** 53
Becket Gremmels

5 **The Training of Clinical Bioethicists: Technical and Human Competencies Through Empathy** 65
David Cerdio-Domínguez and Jonathan McFarland

6 **Clinical Bioethics in Latin America: Current State** 81
María Bernardita Portales

Part III Global Bioethics

7 **Demographic Winter: A Global Comparative Analysis of Regional Trends, Socioeconomic Impacts, and Policy Imperatives** 95
Carlo De La Llata-Gómez and Luis Medina-Gual

8 Global Governance and Public Health 117
Alberto García Gómez and Martina Todaro

9 Beyond Borders: Global Bioethics and the Humanization of Migration Policy 131
María Victoria Fernández Molina and José Rafael Grijalva Eternod

10 Sustainable Development Goals and the Universal Declaration on Bioethics and Human Rights 149
Henk ten Have

11 Bioethics and Armed Conflicts 163
Carlos Alfonso Viesca y Treviño and Maríablanca Ramos R. de Viesca

Part IV Artificial Intelligence and Neuroethics

12 Bioethical Challenges of Regulating Neurorights in the Field of Neuroscience 185
José Damián Carrillo-Ruiz, Selene Márquez-Chaparro, José Rodrigo Carrillo-Márquez, and María Fernanda Carrillo-Márquez

13 Bioethics and Artificial Intelligence 201
María Elizabeth de los Ríos-Uriarte and José Sols Lucia

Part V Bioethics Education

14 Integral Formation in Bioethics and Research Ethics 213
Raúl Jiménez Piña and Aida Cristina Coss Rojas

15 Bioethical Education in Ibero-America: Main Characteristics and Current Challenges 223
Alejandro Sánchez-Guerrero, María Victoria Fernández Molina, and Daniela Casillas-Gómez

16 Medical Simulation in Clinical Bioethics Education 245
Ariana Cerón Apipilhuasco

Part I
Fundamental Bioethics

Chapter 1
Gender Perspective in Bioethics: Toward an Integral Reflection

Jennifer Hincapie Sanchez and Fernanda de Blas López

1.1 Introduction

The gender perspective is defined as an analytical approach that explores the social, cultural, and economic implications associated with gender, determining how these affect different groups in various contexts. This framework is not limited to the recognition of biological differences but emphasizes power dynamics, socially assigned roles, and the structures that support inequalities in personal and professional development and in access to the safeguarding of fundamental rights. Understood as an interpretive tool, the gender perspective seeks to reveal how identities and life experiences are shaped by particular historical contexts. Its main objective is to question and transform discriminatory systems to advance toward equality, social justice, and respect for diversity in all its forms [1].

The academic genesis of this perspective is closely linked to the feminist movements that emerged in the second half of the twentieth century, especially in the 1960s and 1970s. These collective actions demanded equality in various contexts: political, economic, cultural, and social, generating profound questions about traditional practices. A significant moment was the World Conferences on Women organized by the United Nations, held in Mexico City (1975), Copenhagen (1980), Nairobi (1985), and Beijing (1995) [2]. These Conferences consolidated the gender perspective in academic and political analysis, catalyzing a substantial change in public policies at a global level, underlining the imperative need to confront the discrimination and violence specifically exercised against women, and promoting comprehensive approaches that encompass security and equality in education, work, the economy, and politics [3].

J. H. Sanchez (✉) · F. de Blas López
University Bioethics Program UNAM, Mexico City, Mexico
e-mail: jhincapie@unam.mx

D. Cerdio-Domínguez, M. E. de los Ríos-Uriarte (eds.), *Current Bioethics*, New Paradigms in Healthcare, https://doi.org/10.1007/978-3-032-25294-4_1

The growing documentation and visibility of gender-based violence exposed the need for interdisciplinary analyses capable of identifying its causes and proposing effective solutions. Areas of knowledge such as psychology, anthropology, history, philosophy, law, and the health sciences began to incorporate this approach, thus challenging traditional paradigms. This process involved not only a methodological renewal but also a profound historical contextualization, which, by questioning stereotypes, traditional roles, and rigid social norms, became an indispensable tool for fostering profound transformations toward effective and sustainable equality between genders [4].

The gender perspective in bioethics emerges as an essential and critical approach that allows for the identification, analysis, and correction of structural inequalities that have been entrenched in the social, health, and biomedical spheres. Hegemonically, bioethics has been influenced by universal and apparently neutral ethical principles, concealing biases and discrimination linked to gender. The incorporation of the gender perspective highlights these lacks of equality, promoting a broader and contextualized framework for reflective analysis that transcends the principlist model of Beauchamp and Childress [5], which offers an ethical reflection that recognizes and questions unequal power relations. This vision responds to a growing critique of traditional models that systematically ignored the diversity of human beings and their life experiences, thereby marginalizing the specific needs of female and diverse bodies, thus limiting the possibility of achieving real equality in the health field.

In this context, international organizations such as UN Women emphasize that gender equality should be understood as a cross-cutting objective for achieving sustainable development and for guaranteeing full respect for human rights in all contexts, including health and bioethics [6]. This conceptual framework holds that gender equality implies not only formal equal treatment but also proactive actions that address systemic and structural inequalities, promoting the creation of public policies and social practices that eradicate gender gaps.

Recent studies in bioethics show that a critical perspective on gender allows for the visibility not only of inequalities based on sex-gender difference but also those derived from gender and sexual orientation diversities. This approach makes it possible to comprehensively address the needs of the multiple populations that make up a society, thus promoting a reflection that actively values diversity in care and biomedical research [7]. The incorporation of the gender approach has produced a paradigmatic shift in how contemporary bioethical problems are proposed, raised, and resolved, providing a key tool for strengthening justice in health systems globally. International institutions such as the World Health Organization (WHO) and medical ethics research centers like The Hastings Center have emphasized the relevance of its incorporation to strengthen ethical analysis [8].

Feminist theory has critically questioned the dominant paradigms in philosophy, ethics, and science, exposing the embedded gender assumptions in these fields. Thus, this field of knowledge is positioned as a response to the limitations of traditional bioethics, which has frequently privileged individual autonomy over relational contexts. This has led to a neglect in considering the systemic inequalities that condition people's moral agency and health outcomes [9]. Instead of analyzing ethical dilemmas in isolation, feminist bioethicists emphasize the importance of context, power relations, and the social structures that condition moral decisions. Feminist contributions to bioethics include critiques of the "abstract moral agent," a supposedly rational and self-sufficient figure often modeled on male experiences, as well as the proposal of an "ethics of care" that values empathy, interdependence, and vulnerability. These approaches have enriched ethical discourse by making visible the lived experiences of those who have been historically excluded from ethical consideration, particularly women, caregivers, and marginalized populations.

When a person is born, their corporeal sexual identity is poorly defined. It is through processes of education and socialization that the body acquires characteristics that allow it to be identified as masculine or feminine, beyond the specific biological particularities of each individual's corporeality [10]. Furthermore, the concept of gender is used in contrast to that of sex, thus attempting to shift the analysis of power relations between men and women from the realm of biology to that of culture and the symbolic order. That is, the notion of gender underscores that what we understand as men and women are not natural realities, but cultural constructs [11], just as the concept of corporeality is. Both definitions directly impact the lives of women, who experience their body and their gender culturally and socially through pathological experiences.

Gender is not reduced to a biological event; it is a socially constructed concept laden with political meanings that influence how people are perceived, treated, and valued in society. In the context of bioethics, a binary and reductionist view of gender fails to capture the complexity of individual experiences or institutional structures. As Judith Butler [12] argued, gender is performative: it is constituted through repeated social practices that align bodies, identities, and expectations. The shift from binary frameworks to an intersectional analysis has allowed for significant advances in the field. Intersectionality, coined by Kimberlé Crenshaw [13], describes the way in which different systems of oppression—such as sexism, racism, classism, or heteronormativity—interact to generate unique forms of disadvantage. An intersectional perspective in bioethics allows for visualizing, for example, the ways in which an Afro-descendant trans woman with limited economic resources may experience health inequalities very differently from a cisgender Caucasian woman, even within the same health system. Feminism, then, offers a lens that allows health problems to be addressed from another perspective. This common approach highlights key elements such as relational autonomy, which conceives the autonomous subject not as an isolated individual, but as a function of the interrelation and interdependence between people. Similarly, standpoint epistemology maintains that truth and knowledge are always situated and partial, which implies that there are no absolute or universal truths disconnected from context. The ethics of plural dialogue prioritizes responsibility, emphasizes the importance of starting from concrete and contextualized experiences, incorporating comprehensive listening as a method to "put oneself in another's shoes" and understand their demands [14].

The initial development of bioethics in the mid-twentieth century coincided with significant advances in biomedicine, such as organ transplants, research with human subjects, and reproductive technologies. However, much of this early bioethical discourse was shaped by male academics from liberal philosophical traditions, who often privileged abstract reasoning over embodied experience [15]. As a result, the normative frameworks that dominated bioethics—particularly principlism—tended to universalize a model of a moral agent that was implicitly masculine, White, and autonomous. "It is not an exaggeration to say that in the minds of many researchers, the paradigmatic research subject remains, in a way, a mature, respectable, moderately well-educated, clear-minded person who can read and write, a well-positioned and self-sufficient U.S. citizen" [16]. This gender exclusion was not merely theoretical. In practice, it contributed to real disparities in health ethics and policy. For example, many early clinical trials systematically excluded women, which led to medical guidelines based primarily on male physiology and responses [17]. The ethical oversight of these trials rarely questioned the gender implications of such exclusions.

Androcentrism—the practice of centering men and their experiences as the norm—has influenced both the content and the methodology of biomedical research. Historically, women were excluded from clinical research due to concerns about hormonal fluctuations and potential risks during pregnancy. Consequently, many drugs and treatments were tested on men but later prescribed to women without sufficient evidence of their efficacy or safety in female bodies [18]. This androcentric discourse present in biomedicine has had profound repercussions, not only guiding medical interpretations of how women's diseases are described, diagnosed, and treated but also justifying and perpetuating inequality between the sexes through mechanisms such as the medicalization, invisibilization, and psychologization of female physiology and diseases [19]. In bioethical debates, the impact of these practices has frequently been ignored. Issues that uniquely or disproportionately affect women—such as sexual and reproductive health, maternal mortality, or gender-based violence—were often considered peripheral issues or "special topics," rather than central ethical concerns. This way of framing problems implicitly positions male health as the universal standard and female health as a marginal category.

Michel Foucault's concept of the "medical gaze"—the way in which medicine objectifies the patient and abstracts their illness from the social context—has been key in feminist critiques of medical ethics. From this perspective, the clinical encounter is not simply a neutral interaction, but a space where institutional and gendered power dynamics are exercised [20]. Thus, from a Foucauldian perspective, the body is the fundamental site where sexual identity is constituted, thereby becoming a device that symbolically and materially connects all the social and cultural representations through which gender identities are configured [21].

Feminist scholars have highlighted the way in which this gaze disproportionately affects women and people with diverse gender identities, frequently leading to underdiagnosis, misdiagnosis, or the dismissal of symptoms. Conditions such as endometriosis, chronic pain, or autoimmune disorders are often misunderstood or minimized, largely due to gender biases that frame female suffering as exaggerated or psychosomatic [22]. These structural biases are not accidental: they are embedded in the institutions and epistemologies that govern medical knowledge and ethical standards. Addressing them requires a critical review of how bioethics defines moral problems, whose voices are heard, and what values are prioritized. The integration of a gender perspective in bioethics reveals the extent to which gender shapes ethical dilemmas in various areas of health and biomedicine. Far from being neutral or universally experienced, ethical problems are mediated by social, cultural, and institutional structures that frequently exacerbate gender-based inequalities.

The ethics of sexual and reproductive health is one of the areas most relevantly traversed by gender within bioethics, as it has a direct impact on bodies with the capacity to gestate, as well as on the social expectations linked to reproduction. Debates surrounding abortion, contraception, and assisted reproductive technologies (ART) illustrate how women and people with the capacity for pregnancy are often deprived of full bodily and moral autonomy under the pretext of protecting potential life or preserving traditional family structures. Ethical arguments about abortion, for example, often prioritize the rights of the fetus while minimizing the agency of pregnant persons. At the same time, policies that restrict access to reproductive care disproportionately affect marginalized women—particularly those living in poverty, in rural areas, or under oppressive legal regimes [23].

Assisted reproduction introduces additional concerns. While technologies like in vitro fertilization or surrogacy have expanded reproductive options, they also raise ethical questions about commodification, exploitation, and informed consent—especially when women with economic disadvantages are recruited as egg donors or surrogates for wealthier clients [24]. Gender influences not only the availability of health services but also the way decisions are made within medical contexts. Women are often prone to performing caregiving roles, postponing their own medical treatment, and experiencing disparities in the diagnosis and management of pain [25]. Gender norms also affect the degree to which health professionals take a patient's symptoms seriously, which in turn impacts diagnostic accuracy and treatment outcomes.

In patriarchal societies, decision-making may not rest with the person directly affected. Women may need permission from male relatives to access health services or face social stigmatization for seeking care related to their sexual or reproductive health. These dynamics challenge the bioethical principle of autonomy and require a more nuanced and culturally sensitive approach.

The ethics of care, a concept originating in feminist philosophy, holds that ethical reasoning must consider the relational and emotional contexts of human life [26]. Care work, whether performed by nurses, assistants, or unpaid caregivers in the domestic sphere, is overwhelmingly done by women and often remains invisibilized or undervalued in ethical deliberations. Bioethics has historically tended to privilege acute and high-stakes dilemmas, such as organ transplants or end-of-life decisions, over the everyday and sustained ethics of care. A gender perspective insists that the moral value of care work must be recognized, and that policies should support, not exploit, those who provide care. While traditional bioethics often starts from heteronormative assumptions, a gender-sensitive bioethics must also address the ethical issues affecting LGBTIQ+ people. Trans and nonbinary people face specific challenges in accessing affirmative medical care, suffer discrimination in clinical settings, and must navigate consent processes in interventions that affect their bodily integrity [5]. Ethical frameworks must evolve to ensure respect for gender diversity and protect against medical practices that pathologize or marginalize trans identities. This implies rethinking normative assumptions about sex, gender, and the "ideal body" in both medical education and bioethical analysis.

Case-based analysis is an essential tool in bioethics, as it allows theoretical approaches to be contrasted with the complexities of the real world. When analyzed from a gender perspective, many contemporary bioethical controversies reveal the persistence of inequalities, systemic negligence, and the urgent need for inclusive ethical frameworks.

Both in the Global North and South, maternal mortality remains a critical ethical concern—shaped by intersectional factors such as poverty, race, and gender. In the United States, for example, Afro-descendant women are three times more likely to die from pregnancy-related causes than Caucasian women, even when controlling for factors like income and education [27]. These disparities reflect not only failures in clinical care but also broader structural injustices, such as racism, lack of prenatal support, and medical negligence. The reproductive justice movement, led by racialized women, reframes bioethical debates by demanding not only the right to avoid pregnancy (through contraception and abortion) but also the right to have children and raise them in safe and supportive environments [28]. This approach challenges conventional bioethics to expand its scope beyond individual choice, incorporating the social, economic, and environmental determinants of reproductive autonomy.

During the COVID-19 pandemic, various ethical issues arose concerning vaccine distribution, prioritization, and vaccine hesitancy, all with gender dimensions. Women, especially those in frontline care and service jobs, bore a disproportionate burden of exposure and infection, while also facing an increase in domestic violence and the demands of unpaid care [29]. In some contexts, vaccine trials lacked data disaggregated by sex and gender, which limited the ability to evaluate differential effects. Furthermore, public health messages often failed to consider cultural norms or gender barriers to accessing vaccination, especially in conservative or patriarchal societies. These examples demonstrate that even apparently neutral public health strategies can perpetuate gender inequities if they are not informed by an intersectional analysis.

The growing visibility of the rights of trans people has brought under ethical scrutiny the medical frameworks that regulate gender-affirming care. Practices such as mandatory psychological evaluations, restricted access protocols, and limitations on access to hormones or surgeries are often justified as protective measures but can become forms of paternalism or discrimination [30]. Bioethicists have a responsibility to question the medicalization of trans identities when it reinforces normative assumptions about gender and pathologizes difference.

Ethical practice should focus on informed consent, bodily autonomy, and respect for self-affirmed gender identity, especially in the care of minors and young people undergoing gender affirmation processes.

In countries like Canada, Peru, and others, cases of forced or coercive sterilizations of women belonging to Indigenous peoples have come to light, constituting flagrant violations of human rights and bioethical constructs. These practices, often justified under discourses of population control or poverty reduction, reveal the intersection of colonialism, racism, and gender-based violence within medical institutions [31]. The ethical failures in these cases are not merely historical; they persist in subtler forms through unequal access to reproductive services, cultural insensitivity in care, and the erosion of informed consent. Addressing these injustices requires a decolonial and feminist bioethics that listens to community voices and promotes reparative justice.

Transforming bioethics into a more inclusive and equitable discipline requires deliberate efforts to integrate gender awareness into both theory and practice. An inclusive bioethics with a gender perspective is not a specialized or marginal field, but a fundamental reorientation of ethical inquiry that recognizes how social identities shape experiences of health, illness, and care. A bioethics with a gender focus is enriched by a plurality of methods that challenge the limitations of conventional analytical models, which are frequently rooted in abstraction and supposed neutrality. These alternative frameworks privilege experience, context, and power relations, offering a more situated and justice-oriented understanding of ethical dilemmas.

Based on the work of Afro-descendant feminist theorists, intersectional analysis provides a framework for understanding how systems of oppression such as sexism, racism, classism, heteronormativity, and ableism interact to shape complex experiences of vulnerability and privilege. In bioethics, intersectional analysis allows professionals and academics to overcome one-dimensional thinking and recognize how overlapping identities affect access to health, treatment outcomes, and the constitution of the ethical subject [32]. For example, the ethical implications of a public health intervention can vary drastically between a cisgender Caucasian woman and a racialized migrant trans woman, differences that traditional models often fail to capture.

Narrative ethics places the moral importance of the story at its center, emphasizing that ethical understanding is derived not only from abstract principles but also from lived experiences and personal stories—especially from those voices that have been historically marginalized in dominant discourses [33]. This approach rehumanizes bioethics by restoring subjectivity, emotion, and context to ethical deliberation. In clinical settings, narrative ethics invites professionals to listen deeply to patients' stories as a way to understand not only their symptoms but also their identities, relationships, and moral worlds. From a gender perspective, it allows for visualizing how norms of femininity, masculinity, and gender dissent shape narratives of illness and moral agency.

Community-Based Participatory Research (CBPR) is a collaborative approach to knowledge production that positions communities not only as objects and subjects of study but also as co-creators of knowledge and ethical reflection. This methodology is especially powerful in contexts where historically oppressed groups have been objectified, silenced, or exploited in the name of science. In an inclusive bioethics with a gender perspective, CBPR fosters ethical reflection based on dialogue, mutual respect, and shared authority [34]. It challenges researchers to be accountable to the communities they study and to design interventions that are culturally relevant, socially just, and sensitive to the lived realities of those who need them most. These approaches question the idea of an "objective" and detached bioethics, advocating instead for pluralism, cultural competence, and epistemic justice.

Advancing toward an inclusive bioethics with a gender perspective requires not only theoretical transformations but also concrete changes in institutional structures and public policies. Ethical principles, if not accompanied by structural reforms, risk remaining at the aspirational level. Institutions, from hospital units and research centers to funding agencies and educational programs, play a decisive role in both the reproduction and the overcoming of gender inequalities. One of the fundamental reforms consists of the systematic collection, analysis, and publication of data segregated by sex and gender. For a long time, biomedical research has operated under the assumption that male physiology can serve as a universal standard, which has led to the underrepresentation of women and the invisibilization of intersex and trans experiences [17]. Institutional policies must demand inclusive recruitment in clinical trials to ensure the reporting of gender differences in outcomes and encourage analyses that transcend binarism.

International standards such as the SAGER (Sex and Gender Equity in Research) guidelines offer concrete recommendations for incorporating gender analysis in scientific publication and should be adopted by journals, research ethics committees, and funding bodies [35].

Reform must also begin in bioethics and medical education. Medical curricula often omit the sociocultural dimensions of gender or the structural factors that influence patients' experiences. This leads to clinical practices that reproduce biases, such as the underdiagnosis of pain in women or the assumption of heteronormative family structures in treatment planning [25]. Institutions must integrate critical gender studies, intersectional theory, and cultural diversity training into professional health education. These contents should not be treated as elective or peripheral subjects, but as essential components of ethical competence. Creating inclusive institutions in terms of gender also requires internal reforms in hiring, leadership, and governance processes. Women and people with diverse gender identities continue to be underrepresented in decision-making roles in academic bioethics and health administration. Fostering equitable representation on ethics committees, editorial boards, and regulatory bodies enhances the legitimacy and responsiveness of institutional structures [36].

Institutions must develop accountability mechanisms to ensure that gender equity is not just a discursive commitment. This includes measurable goals, transparent evaluation criteria, and the active participation of affected communities in oversight processes. Establishing offices of diversity, equity, and inclusion with real authority can act as a structural safeguard against gender discrimination and epistemic exclusion. On a broader scale, national governments and international organizations must adopt gender mainstreaming as an ethical imperative in all health-related policy. This implies evaluating how laws, public health programs, and resource allocation have a differential impact based on gender and addressing the systemic disadvantages that result. For example, maternal health policies must align with reproductive justice frameworks, and pandemic preparedness plans must consider the gendered impacts on economic precarity and care burdens [29].

Advancing toward an inclusive bioethics with a gender perspective implies reimagining the field as a social justice project. Instead of limiting itself to reacting to technological innovations or individual cases, bioethics can play a proactive role in identifying and dismantling structural barriers that affect health and dignity. This reformulation is in tune with calls for a "critical bioethics": a bioethics that interrogates power structures, adopts an active stance, and works in solidarity with movements for health equity, reproductive justice, the rights of people with disabilities, and LGBTIQ+ inclusion [37]. In this framework, gender equality is not an ethical add-on, but a central moral demand.

While the integration of a gender perspective in bioethics is increasingly recognized as necessary, its implementation faces both conceptual and institutional challenges. At the same time, this evolving field presents valuable opportunities to enrich ethical analysis, promote justice, and advance toward more inclusive health systems. One of the most persistent obstacles to an inclusive bioethics with a gender perspective is institutional resistance. Despite growing evidence of gender inequalities in health outcomes and medical care, bioethics continues to be heavily influenced by dominant frameworks that prioritize a supposed abstract universality over social context. This resistance is often reinforced by forms of epistemic injustice—specifically the testimonial and hermeneutical forms—in which the knowledge and experiences of women and people with diverse gender identities are dismissed or misinterpreted [38].

For example, patients who report pain, fatigue, or psychological distress are sometimes dismissed due to gendered assumptions about emotionality or exaggeration, which leads to erroneous diagnoses or deficient treatments. These situations trigger not only clinical but also ethical failures, rooted in the devaluation of certain voices within the medical discourse.

Efforts to incorporate a gender approach in bioethics must also confront global inequalities and cultural specificity. What is understood by gender justice can vary significantly according to social, legal, and religious contexts. For example, advocating for reproductive rights in a liberal democracy may differ profoundly from doing so in a society with entrenched patriarchal or theocratic norms. Bioethicists must navigate these tensions carefully, avoiding both cultural imperialism and moral relativism. An inclusive bioethics with a gender approach must be dialogical and context-sensitive, committed to local communities while upholding universal principles such as human dignity, substantive equality, and informed consent.

Despite these challenges, the growing attention to gender in bioethics offers unique opportunities. It expands what is considered ethically significant by making visible hidden labors (e.g., unpaid care), revealing ignored injustices (such as underrepresentation in research), and introducing new moral subjects (such as trans youth, surrogate mothers, and intersex people) into the ethical discourse. This shift not only increases ethical sensitivity but also expands the ethical ideas, challenging bioethics to address issues of identity, corporeality, and relationality in greater depth.

A bioethics informed by a gender approach benefits from interdisciplinary work. Contributions from sociology, anthropology, law, gender studies, and public health are essential for building more robust ethical frameworks. These collaborations can help translate theoretical critiques into practical changes within medical education, healthcare delivery, and public policies. Academic institutions, professional associations, and economic funding bodies have a key role in fostering this interdisciplinary work. The opportunity lies in creating networks of academic and professional communities committed to a bioethics that is inclusive, reflective, and transformative.

The integration of a gender perspective in bioethics has traditionally focused on clinical care, reproductive health, and biomedical research, areas where power dynamics and access related to gender are particularly visible. However, to realize the transformative potential of feminist and intersectional ethics, it is necessary to extend this analysis to emerging and underexplored areas of bioethical reflection. In doing so, it is revealed that gender is relevant not only for the analysis of patient-centered care but also for understanding environmental responsibilities, the design of technologies, and the organization of global health responses. For example, environmental bioethics examines the moral relationships between human beings, non-human sentient beings, and ecological systems. A gender lens reveals how environmental degradation disproportionately affects women and populations with diverse gender identities due to traditional social roles, economic marginalization, and scarce political representation.

Another example is that women are often responsible for securing the provision of water, food, and fuel in rural and peri-urban communities. When environmental resources are depleted or contaminated due to climate change, mining, or industrial waste, it is they who bear the brunt of the immediate consequences. These gender vulnerabilities are exacerbated by limited access to land rights, environmental decision-making spaces, and climate financing [39]. A feminist environmental bioethics advocates for a relational and justice-oriented ethics that recognizes care for ecosystems as inseparable from care for vulnerable human communities.

The rapid incorporation of artificial intelligence (AI), big data, and digital health technologies into medical systems presents novel ethical challenges, often traversed by gender, though not always recognized as such. Machine learning models trained on datasets that reflect historical gender biases—such as the underdiagnosis of cardiac diseases in women, the lower representation of feminized bodies in clinical research, or racialized misclassification in dermatological applications—can replicate and even amplify these injustices in supposedly "neutral" technologies [40–42]. For example, one study showed that algorithms used to allocate healthcare resources in the United States systematically underestimated the need for care in Afro-descendant patients, partly because the model used past healthcare costs as a proxy for medical need, an indicator biased by structural barriers to access [40]. Similarly, computer vision–based dermatological diagnosis applications have shown significantly lower performance in people with darker skin tones, due to the scarce representation of these tones in training datasets [43]. These gender and race biases are not mere technical errors: they are a reflection of social structures that have historically shaped both the production of medical knowledge and the distribution of care.

In addition, digital technologies such as reproductive tracking applications, facial recognition, and predictive analysis systems can impose new forms of surveillance and control over gendered bodies. These technologies, when adopted in institutional, labor, or legal contexts without adequate regulation, can expose users to invasive or coercive practices, as has occurred in contexts where menstrual data have been used for legal or commercial purposes [44]. In the case of facial recognition, research shows that error rates in racialized women are much higher, which entails specific risks in policing or migratory contexts [45]. Without a gender-sensitive ethical framework, the development and use of these technologies can deepen inequalities under the pretext of innovation. A feminist bioethics of technology demands that ethical analysis consider not only the outcomes for users but also the design processes, data sources, and systemic power dynamics that shape digital infrastructures. The politics of knowledge production is central to feminist epistemology. In scientific research, the questions that are asked, who receives funding, what methods are privileged, and what results are published, are pro cesses traversed by gender. Women and researchers belonging to minoritized groups frequently face obstacles in academic publishing, institutional recognition, and access to decision-making positions [46, 47]. Meanwhile, studies focused on female health, menstruation, menopause, or diverse gender experiences remain underfunded, under-cited, and poorly integrated into public health policies or emerging technologies [48, 49]. The invisibilization of these areas is not just an omission: it represents a form of epistemic violence that reproduces structural inequalities in the generation of knowledge and its translation into technological applications.

A gender-sensitive research ethics demands more than inclusion: it calls for a structural transformation. This implies recognizing lived experience as a valid source of knowledge, adopting community-led research practices, and questioning the epistemic authority of traditional institutions that have historically excluded nondominant perspectives. Gender equality is fundamental in global bioethical conceptualization, especially in contexts of resource distribution, humanitarian crises, and global health governance. Disasters, conflicts, and pandemics tend to magnify existing gender vulnerabilities. A global bioethics with a gender perspective focuses on structural injustices and promotes ethical approaches based on solidarity and egalitarian health policies [50].

In this area, gender inequality intersects with colonial legacies, global capitalism, and geopolitical imbalances. Issues such as vaccine distribution, access to essential medicines, and public health infrastructures often reflect and reinforce global hierarchies in which women's bodies and lives are systematically deprioritized. In these contexts, women face greater risks of sexual violence, loss of income, and reduced access to health services. However, their voices are frequently absent from decision-making spaces. A gender-sensitive global bioethics aligns with decolonial and postcolonial frameworks that seek to redistribute epistemic and material power, advocate for reproductive and environmental justice, and promote global solidarity based on substantive equality.

1.2 Conclusions

The integration of a gender perspective in bioethics is not simply an ethical preference, nor does it respond to a trending topic; it is a moral necessity. A bioethics that does not address the issue of gender runs the risk of perpetuating injustice, excluding critical voices, and reinforcing structural inequalities. From reproductive health to environmental degradation, from medical AI to the ethics of scientific research, gender operates not only as a background variable but also as a central axis of ethical concern. An approach to bioethics with a gender perspective demands going beyond abstract principles to encompass lived experiences, the social context, and intersectional analysis. It requires rethinking fundamental concepts—such as autonomy, justice, and beneficence—in a way that reflects the diversity of human lives and the complexity of moral agency. A transformative and gender-inclusive bioethics must go further: it must align with broader movements for social and epistemic justice. This includes challenging the hierarchies of knowledge, resisting institutional inertia, and fostering collaborative practices that prioritize the voices of the people most affected by health and environmental disparities, technological exclusion, and ethical negligence.

The future of bioethics lies in its capacity to evolve, not only as a field of academic research but also as a tool for collective care, political advocacy, and ethical transformation. A bioethics with a gender perspective and committed to justice is better prepared to respond to the urgent and emerging dilemmas of our time.

References

1. Scott JW. Gender: a useful category of historical analysis. Am Hist Rev. 1986;91(5):1053–75.
2. Conferencias mundiales sobre la mujer [Internet]. ONU Mujeres. [cited 2025 May 5]. Available from: https://www.unwomen.org/es/how-we-work/intergovernmental-support/world-conferences-on-women.
3. ONU Mujeres. Declaración y Plataforma de Acción de Beijing. Nueva York: Naciones Unidas; 1995 [citado 2025 May 5].
4. Butler J. Undoing gender. New York: Routledge; 2004.
5. Beauchamp TL, Childress JF. Principles of biomedical ethics. 8th ed. New York: Oxford University Press; 2019.
6. Mainetti MM, Echeverría J, Bernal M, La Rocca S. Educación bioética con perspectiva de género: nuevos derechos en salud. Bios Pap. 2022;1(2):1–10. Available from: https://revistas.unbosque.edu.co/index.php/Bios-Papers/article/view/3899.
7. Ruiz Cantero MT. Perspectiva de género en medicina. Universidad de Alicante; 2019.
8. WHO. Gender mainstreaming for health managers: a practical approach. Geneva: World Health Organization; 2020.
9. Kittay EF. Love's labor: essays on women, equality and dependency. New York: Routledge; 1999.
10. Fernández O. Fenomenología del cuerpo femenino. Investig Fenomenol. 2010;2:243–52.

11. Serret E, Méndez J. Sexo, género y feminismo. México: Suprema Corte de Justicia de la Nación; 2011.
12. Butler J. Gender trouble: feminism and the subversion of identity. New York: Routledge; 1990.
13. Crenshaw K. Demarginalizing the intersection of race and sex: a Black feminist critique of antidiscrimination doctrine. Univ Chic Leg Forum. 1989;1989(1):139–67.
14. Feito L. Aspectos filosóficos de la relación entre las mujeres y la bioética: hacia una perspectiva global. En: Mujer, mujeres y bioética. España: Universidad Comillas; 2010. p. 19–60.
15. Warren V. Feminist directions in medical ethics. In: Kuhse H, Singer P, editors. A companion to bioethics. Oxford: Blackwell; 1996. p. 256–67.
16. Kipnis K. Vulnerability in research subjects: a bioethical taxonomy. In: National Bioethics Advisory Commission, editor. Ethical and policy issues in research involving human participants. Bethesda: National Bioethics Advisory Commission; 2001. p. G1–12.
17. Mazure CM, Jones DP. Twenty years and still counting: including women as participants and studying sex and gender in biomedical research. BMC Womens Health. 2015;15:94.
18. Holdcroft A. Gender bias in research: how does it affect evidence-based medicine? J R Soc Med. 2007;100(1):2–3.
19. Tosal B. Síndromes en femenino. El discurso médico sobre la fibromialgia. Feminismo/s. 2007;10:79–91.
20. Foucault M. The birth of the clinic: an archaeology of medical perception. Sheridan Smith AM, translator. New York: Pantheon Books; 1973.
21. Posada L. Las mujeres son cuerpo: reflexiones feministas. Investig Fem. 2015;6:108–21.
22. Hoffmann DE, Tarzian AJ. The girl who cried pain: a bias against women in the treatment of pain. J Law Med Ethics. 2001;29(1):13–27.
23. Cook RJ, Dickens BM, Fathalla MF. Reproductive health and human rights: integrating medicine, ethics, and law. Oxford: Oxford University Press; 2003.
24. Dickenson D. Commodification of the human body: the ethical challenge. Health Care Anal. 2002;10(1):5–12.
25. Hamberg K. Gender bias in medicine. Womens Health (Lond). 2008;4(3):237–43.
26. Held V. The ethics of care: personal, political, and global. Oxford: Oxford University Press; 2006.
27. Creanga AA, Berg CJ, Ko JY, Farr SL, Tong VT, Bruce FC, et al. Maternal mortality and morbidity in the United States: where are we now? J Womens Health. 2014;23(1):3–9.
28. Ross LJ, Solinger R. Reproductive justice: an introduction. University of California Press; 2017.
29. Wenham C, Smith J, Morgan R. COVID-19: the gendered impacts of the outbreak. Lancet. 2020;395(10227):846–8.
30. MacKinnon B. Autonomy, informed consent, and the ethics of gender-affirming care. Hast Cent Rep. 2020;50(4):17–23.
31. Stote K. An act of genocide: colonialism and the sterilization of Aboriginal women. Fernwood Publishing; 2015.
32. Hankivsky O. Intersectionality 101. Burnaby: Institute for Intersectionality Research and Policy, Simon Fraser University; 2014.
33. Charon R. Narrative medicine: honoring the stories of illness. Oxford: Oxford University Press; 2006.
34. Israel BA, Eng E, Schulz AJ, Parker EA, editors. Methods in community-based participatory research for health. 2nd ed. San Francisco: Jossey-Bass; 2010.
35. Heidari S, Babor TF, De Castro P, Tort S, Curno M. Sex and gender equity in research: rationale for the SAGER guidelines and recommended use. Res Integr Peer Rev. 2016;1(2):1–9.
36. Kuhlmann E, Ovseiko PV, Kurmeyer C, Gutiérrez-Lobos K, Steinböck S, von Knorring M, Brommels M. Closing the gender leadership gap: a multi-centre cross-country comparison of women in management and leadership in academic health centres in the European Union. Hum Resour Health. 2017;15(2):1–7.

37. Rajczi A. Why bioethics should become a social justice project. Hast Cent Rep. 2020;50(5):9–11.
38. Fricker M. Epistemic injustice: power and the ethics of knowing. New York: Oxford University Press; 2007.
39. MacGregor S. Gender and climate change: from impacts to discourses. J Indian Ocean Reg. 2010;6(2):223–38.
40. Obermeyer Z, Powers B, Vogeli C, Mullainathan S. Dissecting racial bias in an algorithm used to manage the health of populations. Science. 2019;366(6464):447–53.
41. Larrazabal AJ, Nieto N, Peterson V, Milone DH, Ferrante E. Gender imbalance in medical imaging datasets produces biased classifiers for computer-aided diagnosis. Proc Natl Acad Sci USA. 2020;117(23):12592–4.
42. Vyas DA, Eisenstein LG, Jones DS. Hidden in plain sight—reconsidering the use of race correction in clinical algorithms. N Engl J Med. 2020;383(9):874–82.
43. Adamson AS, Smith A. Machine learning and health care disparities in dermatology. JAMA Dermatol. 2018;154(11):1247–8.
44. Fox S, Vora P. Reproductive surveillance: privacy risks of fertility and period tracking apps. Technol Soc Mag IEEE. 2021;40(2):10–7.
45. Buolamwini J, Gebru T. Gender shades: intersectional accuracy disparities in commercial gender classification. Proc Conf Fairness Account Transparency. 2018;81:77–91.
46. Witteman HO, Hendricks M, Straus S, Tannenbaum C. Are gender gaps due to evaluations of the applicant or the science? A natural experiment at a national funding agency. Lancet. 2019;393(10171):531–40.
47. Shannon G, Jansen M, Williams K, et al. Gender equality in science, medicine, and global health: where are we at and why does it matter? Lancet. 2019;393(10171):560–9.
48. Criado Perez C. Invisible women: data bias in a world designed for men. New York: Abrams Press; 2019.
49. Nelson A. The social life of DNA: race, reparations, and reconciliation after the genome. Boston: Beacon Press; 2016.
50. Benatar SR. Health care reform and the crisis of HIV and AIDS in South Africa. N Engl J Med. 2005;352(8):786–92.

Chapter 2
The Right to Know: Fertility Awareness and Bioethical Decision-Making

José Alberto Castilla Barajas, María Fernanda Delgado Domínguez, Fernanda Pérez Rea Vela, Claudia Lerma, Abel Lerma, José Marcos Félix Castro, and Ana Lucila Villa Viruette

2.1 Introduction

2.1.1 The Ethical Foundations of Fertility

Bioethics started taking place in the 1970s as a response to growing ethical concerns around scientific and medical advances, especially in those linked to human life, sexuality, reproduction, and dignity. This was first introduced by Van Rensselaer Potter; the term *bioethics* was conceived as a bridge between biological knowledge and moral responsibility [1]. Concurrently, André Hellegers' work at Georgetown

J. A. C. Barajas (✉) · M. F. D. Domínguez
Facultad de Bioética, Universidad Anáhuac México, Mexico City, Mexico
e-mail: alberto.castilla@anahuac.mx

F. P. R. Vela
Universidad Anáhuac México, Mexico City, Mexico

C. Lerma
Departamento de Biología Molecular, Instituto Nacional de Cardiología Ignacio Chávez, Centro de Investigación en Ciencias de la Salud (CICSA), Facultad de Ciencias de la Salud, Universidad Anáhuac México, Mexico City, Mexico

A. Lerma
Instituto de Ciencias de la Salud Área Académica de Psicología, Universidad Autónoma del Estado de Hidalgo (UAEH), Pachuca, Mexico

J. M. F. Castro
Oncología Pediátrica, Centro Médico Nacional la Raza IMSS, Hospital Pediátrico Moctezuma de la CDMX, Facultad de Ciencias de la Salud Universidad Anáhuac México, Mexico City, Mexico

A. L. V. Viruette
Centro de Lenguas, Universidad Anáhuac México, Mexico City, Mexico
e-mail: ana.villav@anahuac.mx

D. Cerdio-Domínguez, M. E. de los Ríos-Uriarte (eds.), *Current Bioethics*, New Paradigms in Healthcare, https://doi.org/10.1007/978-3-032-25294-4_2

University contributed significantly to institutionalizing bioethics in the biomedical field. From its origins, bioethics has been intrinsically linked to the ethical reflection on sexuality and fertility, recognizing the decisions related to procreation, identity, responsibility, and freedom [1].

Fertility awareness (FA) fits into the ethical picture quite well. Fertility awareness refers not only to the ability to identify fertile and infertile periods within the menstrual cycle but also a broader consciousness of one's reproductive health, rooted in both biological understanding and personal freedom. The term *fertility* is deeply human because it is mediated by *consciousness, decision-making, and values*. From a personalist bioethical perspective, fertility must be understood not only as a biological capacity but also as a dimension of the human person, linked to freedom and responsibility. To choose whether to have children or not, and under which circumstances, it involves a deep reflection, intention, and choice.

Sexual and reproductive health is now widely recognized as a fundamental component of human well-being, and it is an important basic human right. Despite actual advances in public health and policies and an increasing access to health-related information, there are significant gaps in young women's understanding, management, and awareness of their own fertility. In this sense, promoting fertility awareness among young women is not only an educational or public health objective but also a bioethical matter. It helps women to practice their reproductive autonomy in an informed and ethical manner, in harmony with their values, body integrity, and life plans. The lack of fertility awareness, particularly in environments where education is distorted by cultural, social, or ideological pressures, represents vulnerability and ethical injustice [2].

Understanding fertility awareness through the lens of bioethics invites us to make a deep reflection on the essential role of freedom in reproductive decision-making. True freedom is rooted in reason and oriented toward the good. It also involves the capacity to choose, evaluate our options, and to assume big responsibilities for the consequences of our own actions. Fertility awareness is a space of personal growth. It has a mature understanding of the body as a meaningful part of oneself, not as an object to be controlled, but as a subject of care and respect. By having a bigger awareness on fertility, women are not only gaining information, but they are also cultivating a relationship with their body, with their capacity to bring life into the world (or not), and with their *right to decide* responsibly.

2.1.2 Defining Infertility: A Global Concern

Infertility is a medical condition that is defined as the inability to conceive after 1 year of unprotected sexual intercourse with the intention of achieving pregnancy [3]. It is a worldwide concern that affects a significant part of the reproductive-age population. According to the World Health Organization (WHO), it is estimated that approximately 10%–15% of couples of reproductive age globally experience difficulties in conceiving [3]. This figure varies by region and may be influenced by sociodemographic and economic factors [4].

2.1.2.1 The Psychosocial Impact of Infertility

Infertility can have a profound impact on the physical and emotional health of affected individuals. Physically, it may be linked with underlying medical conditions such as hormonal disorders, gynecological diseases, or general medical conditions like obesity or insulin resistance can affect fertility. Additionally, the process of trying to conceive, the diagnosis and treatment of infertility can be emotionally difficult and stressful [5]. Emotionally, infertility can cause feelings like frustration, sadness, anxiety, and stress. It can also have a significant impact on self-esteem and the perception of one's identity as an individual or as part of a couple. These emotional aspects may extend further, affecting the overall quality of life and general well-being of those involved [5]. Therefore, addressing infertility from a personalist bioethical perspective is essential to truly understand and manage this health issue, considering not only the medical aspects but also the values, beliefs, and reproductive autonomy of those involved.

2.1.3 Fertility Awareness and Reproductive Autonomy

Throughout a woman's life, reproductive health decision-making should be guided by a clear understanding of autonomy and core bioethical principles. These considerations are central to the concept of fertility awareness, which goes beyond mere knowledge of biological reproductive processes. It also includes the individual's attitudes, beliefs, and understanding of their own sexuality and reproductive potential [6].

Reproductive autonomy refers to the right of individuals to make *free and informed decisions* about their sexual and reproductive health. This bioethical principle acknowledges both the capacity and the right of individuals to make choices that directly affect their reproductive lives. However, in practice, reproductive autonomy can be influenced by different factors, including cultural norms, social pressures, legal restrictions, and limited access to reproductive health resources and services. These barriers can result in situations where individuals do not have full freedom to make informed decisions that align with their personal reproductive needs and desires.

Fertility awareness goes beyond the basic understanding of the biological processes of reproduction. It also includes the ability to make informed and autonomous decisions about an individual's reproductive health. This includes access to accurate and comprehensible information on contraceptive methods, family planning, and assisted reproductive options, as well as the ability to evaluate and choose the options that best suit their individual circumstances [7]. It is important to mention that promoting fertility awareness and reproductive autonomy not only benefits individuals in a reproductive age but also plays a crucial role in advancing overall sexual and reproductive health. It helps foster a more informed and empowered society when it comes to reproductive health matters.

Fertility awareness as a structured concept began to take place in the early 2000s, through the pioneering work of Dr. Claudia Lampic. Her study, "*Fertility awareness, intentions concerning childbearing, and attitudes towards parenthood among female and male academics*," marked a turning point in how reproductive knowledge, awareness, and intentions were evaluated in young adults. Prior to her contribution, fertility was often discussed either strictly in biomedical terms or within the context of population health; it rarely integrated psychosocial factors, personal autonomy, or future life planning as part of a broader ethical analysis. Lampic's study was one of the first to empirically investigate the gap between individuals' reproductive intentions and their actual understanding of fertility limitations, particularly related to female age and its relationship with the decrease in their fertility. Her research focused on young, educated populations (male and female university students). the study concluded that there was a significant lack of awareness regarding age-related fertility decline. Many participants intended to postpone parenthood until their mid-30s or later, underestimating the biological consequences it has on their fertility and overestimating the success rates of assisted reproductive health. The dissonance between intention and biological reality is at the center of what Dr. Claudia Lampic defined it as a deficiency in fertility awareness, it is not only lack of information, it goes as deep as a marked disconnection between knowledge, decision-making, and ethical responsibility. Lampic's work highlighted the urgent need for targeted educational strategies that go beyond contraception and promote comprehensive reproductive knowledge [6].

2.1.3.1 The Swedish Fertility Awareness Questionnaire: A Global Standard

Equally important in Dr. Claudia Lampic's work was the development of a structured instrument within her study. *The Swedish Fertility Awareness Questionnaire* was one of the first truly validated tools that was designed to correctly assess fertility awareness in populations. It was designed to assess different dimensions of reproductive intentions, attitudes toward parenthood, and awareness of fertility-related issues. This instrument is now recognized and cited globally in subsequent international research [6].

2.1.3.1.1 Theoretical and Clinical Foundations

The development of the questionnaire emerged from a clear recognition of a growing social phenomenon, the postponement of parenthood in high-income counties, particularly among university-educated populations. Despite broad access to reproductive health services and educational campaigns focused on contraception, there is a significant void in understanding how well-informed young adults truly are about the limits of female fertility, especially as it relates to age. The questionnaire was based on existing empirical literature on reproductive behaviors and intentions, and it incorporated clinical expertise from gynecology, psychology, and midwifery. This interdisciplinary collaboration guaranteed that the instrument would be both medically accurate and psychosocially sensitive.

2.1.3.1.2 Development and Pilot Testing

Before finalizing the instrument, the authors conducted three pilot studies involving medical students, nursing students, and students from different academic programs. A total of 60 participants were involved in these pilot phases. These initial test runs served multiple purposes; they assessed face validity, guaranteed clarity and cultural appropriateness of the language, and provided data for test–retest reliability. Feedback from the participants was used to refine item phrasing and sequencing. The final version of the questionnaire demonstrated satisfactory internal consistency across its main thematic sections, with alpha values of 0.75–0.77 for combined measures, indicating solid reliability.

2.1.3.1.3 Structural Composition of the Instrument

The Swedish Fertility Awareness Questionnaire (SFAQ) consists of 56 items, divided into seven different sections, each designed to evaluate a specific aspect of fertility awareness and related psychosocial constructs.

Demographic Information (10 Items)
This section collects basic but important data on the participants' backgrounds:

- Age
- Type of housing
- Country of origin and parental birthplace
- Relationship status
- Personal or partner experience with pregnancy, miscarriage, abortion, or infertility

These data help on comparative analysis across subgroups and help contextualize responses to subsequent items.

Intentions Regarding Childbearing (4 Items)
This other section explores future reproductive plans from participants.

- An initial dichotomous question—"Do you plan to have children?" (Yes/No)
- Three questions follow for affirmative respondents:
  - Desired number of children
  - Preferred age for first child
  - Preferred age for last child

These questions helped researchers to identify postponement patterns and compare intentions against actual fertile windows.

Importance of Parenthood (1 Item)
Using a visual analogue scale (VAS) that goes from 0 to 100, this single item measure assesses how important it is for the respondent to have children, ranging from *not at all important* to *extremely important*. This item serves as a key indicator of the personal meaning attached to parenthood.

Behavioral Intentions in Case of Infertility (3 Items)
This other section explores respondents' hypothetical reactions to infertility. Items include likelihood (rated on a VAS from 0 to 100)

- Undergoing in vitro fertilization (IVF)
- Choosing adoption
- Accepting life without children

Conditions of Importance for Deciding to Become a Parent (13 Items)
This part explores the perceived importance of different life circumstances in the decision to have children. Respondents rate each condition using a 5-point Likert scale, the alternatives are: Unimportant, Not very important, Rather important, Important, Very important, and No opinion.

Some examples of the evaluated conditions include:

- Having a stable relationship
- Completing higher education
- Having a good financial situation
- Access to childcare
- Career stability and compatibility with parenting

With an alpha value of 0.77, this section reflects internal consistency, and it is very important for mapping the social and economic prerequisites individuals associate with responsible parenthood.

Perceived Life Changes Associated with Parenthood (16 Items)
This section captures anticipated psychological, social, and practical consequences of becoming a parent. Items are rated on a 5-point Likert scale, the alternatives are: Disagree, Partially agree, Mainly agree, Strongly agree, Entirely agree, and No opinion. Some examples include

- Development as a person
- Reduction in career opportunities
- Greater emotional fulfillment
- Perception of decreased freedom
- Feeling "completeness"

This section has an alpha value of 0.75; this highlights its reliability as a measure of perceived parental identity formation and life restructuring.

Fertility Awareness

The last section of the instrument directly assesses participants' knowledge regarding fertility; these questions do have a correct answer and some of them include:

- Age of peak female fertility
- Age when there is a slight and marked decline in female fertility
- Probability of pregnancy at ovulation (in different ages)
- Average success rate of IVF

The significance of the Fertility Awareness Questionnaire extends beyond its methodological strengths. By integrating knowledge, intentions, values, and sociodemographic context, it offers a comprehensive perspective on fertility decision-making in young adults. The tool is adaptable across cultural contexts and has been used in follow-up studies. Additionally, the instrument aligns with bioethical priorities: autonomy, informed consent, justice in access to fertility education, and the right to make responsible reproductive decisions [6].

2.1.3.1.4 Cross-Cultural Adaptations of the SFAQ

United States

Building upon the foundational work of Lampic, the Fertility Awareness Questionnaire (SFAQ) was adapted for use in the United States by Dr. Peterson in 2012 to assess fertility knowledge and parenting attitudes among American undergraduate students. This adaptation aimed to evaluate the extent to which young adults in the United States are aware of age-related fertility decline and their attitudes toward parenthood. Peterson and his colleagues conducted a cross-sectional study involving 246 undergraduate students (138 females and 108 males) from a university in the United States. Participants' mean age was 20.4 years; they were evenly distributed across freshman to senior classes. The adapted SFAQ was administered inline, maintaining the core components of the original instrument [8].

Some of the key findings in Peterson's research was that most participants expressed a desire to have children in the future and nearly 90% indicated that parenthood was an important aspect of their life. Most intended to have their first and last child within what is considered the biologically optimal fertile window. However, despite these intentions, participants demonstrated significant gaps in fertility knowledge. They vastly overestimated the age at which female fertility begins to decline and the success rates of assisted reproductive technologies like in vitro fertilization. For example, many believed that a woman's fertility remains high well into her 40s and that IVF success rates are considerably higher than statistical realities. These misconceptions suggest that young adults may delay childbearing based on inaccurate perceptions of fertility, potentially leading to challenges in achieving desired family size or experiencing involuntary childlessness [8].

Argentina

Jurkowski, et al. in 2021 made a cross-sectional study that explores fertility awareness, parenting attitudes, and knowledge about Assisted Reproductive Technology among Argentinian university students. The researchers explored how well-informed students are about human reproduction, their intentions regarding parenthood, and their understanding of fertility treatments, especially in a cultural context with evolving reproductive legislation and delayed parenthood trends. The study used 680 Argentinian university students (83.2% female with a mean age of 24.7 years), who were recruited via social media through snowball sampling in June 2019. They also used an adapted version of the Swedish Fertility Awareness Questionnaire [9].

Some of their key findings included that 68.6% believed that they had high fertility knowledge, but major misconceptions existed. In total, 78.2% thought that pregnancy was possible at any time during the cycle, only 21.8% correctly identified the fertile window and its duration. In total, 36.2% believed that female fertility declines between ages 45 and 50, clearly underestimating the biological reality, meanwhile 57.5% believed that male fertility never declines. As for attitudes toward parenting, 60.2% of participants without children planned to become parents. The majority (61.4%) intended to have two children, mainly before the age of 35. Financial stability (58.4%), professional goals (55.1%), and finding a suitable partner (43.6%) were seen as the main barriers in becoming a parent [9].

Hong Kong

This cross-cultural study examined fertility awareness and attitudes toward parenthood among Chinese university students in Hong Kong, comparing their responses with similar samples in Sweden and the United States. Despite Hong Kong's modern, highly educated society, the fertility rate remains among the lowest in the world. The study found that Hong Kong students have significant knowledge gaps regarding age-related fertility decline and overestimated the effectiveness of fertility treatments. However, they were less inclined to marry or have children and less likely to seek help for infertility. Cultural and systemic factors, such as traditional family roles, stigma, and a lack of formal sex education, emerged as important influences. This was an online survey made in 2013, using 67 university students in Hong Kong (75% female with a mean age of 23 years). The authors used a Chinese translation and adaptation of the Swedish Fertility Awareness Questionnaire [10].

Key findings regarding fertility awareness included that 76% of their population overestimated the optimal fertility age, 92% underestimated the onset of fertility decline before the age of 30, and 66% overestimated the success rate of IVF. As for the parenthood intentions and attitudes, the key finding included that 20% did not intend to have children. The top concerns in the population regarding the obstacles to parenthood were not finding a suitable partner (68%), career goals (49%), and financial concerns (39%). Compared to Western results, the importance of childbearing scored significantly lower in Hong Kong students; they were also less likely to pursue medical treatment or adoption in case of infertility and were more likely to accept childlessness [10].

Cultural influences play a significant role in shaping perceptions related to fertility in places like Hong Kong. Traditional family expectations and the persistent stigma surrounding infertility often discourage individuals from seeking medical treatment. Furthermore, adoption remains culturally less accepted, primarily due to the strong societal emphasis on biological lineage. The lack of comprehensive sex education exacerbated the issue by fostering misinformation and limited awareness about reproductive health. Gender roles also contribute to this behavior, despite high female participation in the workforce, childbearing continues to be viewed predominantly as a woman's responsibility, a burden that is frequently managed through the employment of domestic helpers. A notable mismatch exists between individuals' reproductive intentions and biological realities; many aspire to delay childbearing without fully considering the limitations of fertility with age, which may result in emotional distress. These findings highlight the need for culturally sensitive fertility education and the importance of addressing workplace–family balance and fighting against reproductive health stigma in policy development [10].

2.1.3.2 A Systematic Review

Ren, et al. in 2023 made a systematic review aimed to comprehensively assess fertility awareness (FA) among university students and its different factors influencing their understanding. The authors synthesized findings from 21 quantitative studies across 18 countries, most of which had a cross-sectional design. The studies were published between 2006 and 2022; they explored how university students, primarily young adults at their peak reproductive age, perceive and understand human fertility, age-related fertility decline, infertility risk factors, and assisted reproductive technologies (ARTs). Overall, the review found that fertility awareness among students was low to moderate, with female and medical students generally showing higher levels of awareness. Despite being aware that age impacts fertility, students commonly overestimated the age at which fertility begins to decline and misunderstood the optimal age of childbearing. Most students also overestimated the success rates of ART, indicating a reliance on medical technology that may not reflect biological limitations [11].

The study made by Lampic in 2006 occupies a central and influential position within the field. It serves as a foundational reference point, from which a wide range of subsequent research studies have developed. Numerous studies made after Lampics' work are closely aligned with fertility awareness. The influence of this work is evident in how frequently it is cited and conceptually connected to key authors such as Peterson, Yu Ren, and Chan. Lampic's work continues to serve as a cornerstone for discussions on fertility intentions and awareness, underscoring its lasting impact on both clinical understanding and public health education in reproductive contexts.

A significant highlight of this review is the central role of the *Swedish Fertility Awareness Questionnaire* developed by Dr. Claudia Lampic, which emerged as the most widely used and influential tool across studies. This instrument was employed in at least 11 of the 21 reviewed studies, making it the most common standardized measure in fertility awareness research. Its widespread use highlights its reliability and cross-cultural adaptability. The frequent use of this tool across different regions shows us its critical role in shaping comparative fertility awareness research globally. The centrality of the Swedish Fertility Awareness Questionnaire is illustrated in the Connected Papers visual map, which is a graphical representation that shows us the pivotal role of this instrument in shaping and connecting global research on fertility awareness, demonstrating its impact not only in terms of frequency of use but also in fostering academic dialogue and continuity across regions and methodologies [11].

This review also emphasizes that factors such as gender, academic major, age, and educational level significantly influence FA. Female students, and especially those in medical or science fields, consistently demonstrated higher awareness levels. However, even among medical students, awareness gaps persisted, often tied to cultural misconceptions or lack of in-depth reproductive education. Alarmingly, most students reported reading about fertility information from social media or informal sources; this suggests an urgent need for structured, evidence-based fertility education.

2.1.4 Critical Review of International Studies: Barriers in Fertility Awareness

While the reviewed international studies consistently demonstrate low-to-moderate levels of fertility awareness among young adults, a deeper analysis reveals that the roots of this lack of awareness vary significantly across cultural contexts. Beyond simple knowledge deficits, structural, cultural, and psychosocial barriers play a key role in shaping reproductive attitudes and misconceptions.

2.1.4.1 Cultural Stigma and Gender Roles

In Hong Kong, cultural stigmas surrounding infertility represent a major barrier. Traditional family expectations and deeply rooted beliefs about biological lineage led to widespread misconceptions. Many participants were not willing to seek help for infertility, as it could bring shame upon their families. Adoption was culturally discouraged, particularly because of the high value placed on bloodline continuity. Moreover, gender roles were particularly rigid: despite women's high participation in the workforce, childbearing was still viewed primarily as a female responsibility, often carried out with the help of domestic workers. These norms tend to create psychological burdens and limited options for reproductive autonomy [10, 11].

2.1.4.2 Misplaced Trust in Technology

In the United States, the barrier was not lack of education, but rather overconfidence in assisted reproductive technologies. Participants generally underestimated the natural decline in fertility with age and overestimated the success rates of in vitro fertilization, believing it to be a reliable fallback option regardless of maternal age. This misplaced trust led to postponement of parenthood without a realistic understanding of biological limitations. The perception that science can *fix everything* reflects a cultural tendency to de-prioritize preventive reproductive education in favor of technological solutions [8, 11].

2.1.4.3 Educational Gaps

The systematic review by Ren et al. showed that most students accessed fertility information from social media or informal sources, even when they are majoring in medicine or health-related subjects. This points to a global problem: even in highly educated populations, formal education on fertility is poorly structured. It is well known about contraception and sexually transmitted infections, but not age-related fertility decline, infertility risk factors, or realistic ART outcomes. In Argentina, although participants believed they had high fertility knowledge, most were unable to correctly identify the fertile window, and they understood fertility decline only after the age of 45, clearly underestimating the biological reality [8–11].

2.1.4.4 Economic and Social Pressures

Multiple studies highlighted how economic instability and professional aspirations delay the decision to become a parent. In Argentina and Hong Kong, participants named financial insecurity and difficulty finding a stable partner as key reasons for postponement. These are not purely personal choices; they reflect broader societal dynamics, precarious labor markets, lack of childcare infrastructure, and gendered career expectations. Fertility awareness must be framed not just as individual knowledge, but as situated within a social environment that constrains real reproductive freedom [8–11].

2.1.5 Fertility Awareness in Mexico: A Missing Conversation

In Mexico, there is currently a noticeable gap in scientific literature regarding fertility awareness within the general population. Unlike counties in Europe, North America, and parts of Asia, where FA has been increasingly studied and measured, Mexico lacks validated, population-specific tools or nationwide studies that study and analyze people's knowledge, perceptions and attitudes toward fertility, infertility risk factors, and reproductive options such as assisted reproductive technologies. This absence is especially significant considering the social, economic, and cultural characteristics of the country, where topics related to sexual and reproductive health are often shaped by strong religious values, family expectations, and educational disparities.

Given these complex cultural dynamics, it becomes essential to measure fertility awareness in a way that reflects the realities and diversity of the Mexican population. Accurate measurement of FA is not only important from a public health perspective, but it also serves as a foundation for designing effective educational interventions. Understanding the current level of awareness would help identify misconceptions, gaps in knowledge, and potential social or structural barriers that could influence individuals' reproductive decision-making, especially in the context of delayed childbearing, declining global fertility rates, and increasing reliance on ART.

From a personalistic bioethical perspective, this issue is even more urgent. Reproductive autonomy, the right to make informed decisions about it, when and how to have children, requires access to accurate, comprehensible information. Without it, our population may unknowingly make uninformed decisions that interfere with their long-term reproductive goals, potentially leading to involuntary childlessness or emotional and financial distress. The ethical principles of beneficence, justice and respect for autonomy demand that health systems, educators, and policymakers empower their population with the right tools to make informed reproductive choices.

To address this gap, our research aims to assess and compare fertility awareness among Mexican university women enrolled in public and private universities. Our central objective is to explore potential differences in fertility knowledge and perceptions between these two educational sectors, which often reflect broader socioeconomic divides in access to health information and reproductive healthcare.

For this purpose, we used the Beaton guidelines for the cross-cultural adaptation and translation of the Swedish Fertility Awareness Questionnaire. These guidelines are a well-established framework in health sciences for ensuring that instruments keep an equivalence across different languages and cultures. These guidelines are important because simple translations can often result in distortions of meaning and a lack of cultural relevance. Beaton's method consists of five systematic stages, each of which is designed to preserve the conceptual and psychometric integrity of the original tool while adapting it for use [12] in a different cultural setting.

2.1.6 Cross-Cultural Adaptation Process of the Questionnaire

Forward Translation: We used two different persons for two independent translations of the original SFAQ (from English into Mexican Spanish). One of the translators had a C1 level in English and was familiar with the medical field and the objectives of the questionnaire, and the second translator has an A2 level in English and has no medical background.

Synthesis of Translation: A third translator helped by synthesizing both translations. They were compared and combined into a single preliminary version. This step allowed us to resolve discrepancies and use the most culturally appropriate word for each item.

Back-Translation: The synthesized Spanish version was then independently translated back into English by two different bilingual translators who had no prior knowledge about the original version and the questionnaire's purpose. This step was crucial because it verified the accuracy of the translated content, and it guaranteed that the adapted items retained the same meaning as in the original version.

Expert-Committee Review: Our expert panel included Dr. Achar, a professional in reproductive health; Ana Lucila Villa, an expert in linguistics; Dr. Lerma, an expert in methodology and scientific investigations; and Dr. Castro, an expert in the medical field with experience in pedagogy. Together, they reviewed all the prior versions of the translated and culturally adapted instrument. The committee examined semantic equivalence, linguistic equivalence, experiential equivalence, and conceptual equivalence. Their goal was to resolve inconsistencies and to develop a pre-final version of the questionnaire tailored to the Mexican cultural context.

Our research team has completed the cultural and linguistic adaptation of the SFAQ for use in the Mexican context. The upcoming step in this process is the *pre-testing phase,* which will involve administering the pre-final version of the adapted instrument to a sample from our target population, university students and young adults in Mexico. This pilot phase will include feedback from the sample population that will help us evaluate how participants will understand the question, highlight ambiguities, and suggest if any terms or concepts feel culturally or contextually inappropriate.

2.1.7 Next Steps: From Adaptation to Implementation

Once the pre-testing phase is complete, we will finalize a version of the FSAQ that is both linguistically accurate and culturally appropriate to use in Mexico. This adaptation process guarantees that the questionnaire is not only a translated tool, but a validated instrument capable of giving reliable, meaningful data that reflect true fertility awareness, free from cultural distortion or linguistic misinterpretation.

It is important to mention that in order to validate the instrument properly, it is essential to make a large-scale sample in Mexico. A massive sample will help us assess the reliability of the tool across the Mexican population. Once this tool is adapted and properly validated, it can be used in large-scale surveys to assess fertility awareness across different regions and states in Mexico. By aligning our work with other international studies that have used the SFAQ before, we also contribute to the possibility of cross-cultural comparisons. This positions Mexico to become part of a broader international dialogue on fertility awareness and reproductive health.

2.2 Conclusions

Fertility awareness is not merely a matter of health education; it is a bioethical imperative grounded in autonomy, informed decision-making, and human dignity. We have established that fertility is not only a biological function but also a deeply personal and ethical dimension of human life, closely tied to identity freedom, and responsibility. The growing global trend of delayed parenthood, combined with widespread misinformation regarding reproductive limitations, underscores the urgent need to promote accurate and culturally relevant fertility education.

From a personalist bioethics perspective, fertility must be understood not merely as a biological capacity, but as a deep human and relational dimension of the person. Personalist bioethics uphold the intrinsic value of every human being and places human dignity at the center of all moral reflection. Reproductive decisions involve not only biological data or utilitarian outcomes but also freedom, intentionality, and moral responsibility. Fertility awareness empowers individuals, particularly women, to relate to their own bodies not as objects to be managed, but as integral parts of their identity. It enables a deeper, more responsible use of autonomy, one that is not detached from ethical reflection, but fully rooted in it. This perspective affirms that true freedom is not mere choice, but informed and responsible self-determination aligned with the good of the person.

The Swedish Fertility Awareness Questionnaire offers an instrument to assess fertility awareness in a comprehensive, multidisciplinary way. Its integration of biological knowledge, psychosocial context, and personal intentions makes it a powerful tool for research and analysis. The adaptation of this instrument to the Mexican context represents a significant step forward to bridging the knowledge gap in reproductive health and fertility.

References

1. ten Have HAMJ. Potter's notion of bioethics. Kennedy Inst Ethics J. 2012;22:59–82.
2. Pedro J, Brandão T, Schmidt L, Costa ME, Martins MV. What do people know about fertility? A systematic review on fertility awareness and its associated factors. Ups J Med Sci. 2018;123:71–81.
3. World Health Organization. Infertility [Internet]. Available from: https://www.who.int/health-topics/infertility#tab=tab_1.
4. Zegers-Hochschild F, Adamson GD, de Mouzon J, Ishihara O, Mansour R, Nygren K, et al. International Committee for Monitoring Assisted Reproductive Technology (ICMART) and the World Health Organization (WHO) revised glossary of ART terminology, 2009. Fertil Steril. 2009;92:1520–4.

5. Domar AD. Impact of psychological factors on dropout rates in insured infertility patients. Fertil Steril. 2004;81:271–3.
6. Lampic C, Svanberg AS, Karlström P, Tydén T. Fertility awareness, intentions concerning childbearing, and attitudes towards parenthood among female and male academics. Hum Reprod. 2006;21:558–64.
7. American College of Obstetricians and Gynecologists. Committee opinion no. 644: the importance of contraception for women's health. Obstet Gynecol. 2015;125:250–5.
8. Peterson BD, Pirritano M, Tucker L, Lampic C. Fertility awareness and parenting attitudes among American male and female undergraduate university students. Hum Reprod. 2012;27:1375–82.
9. Jurkowski L, Manubens R, Ryberg JO, Rossi M. Fertility awareness, attitudes towards parenting, and knowledge about assisted reproductive technology among university students in Argentina. JBRA Assist Reprod. 2021;25:453–8.
10. Chan CHY, Chan THY, Peterson BD, Lampic C, Tam MYJ. Intentions and attitudes towards parenthood and fertility awareness among Chinese university students in Hong Kong: a comparison with Western samples. Hum Reprod. 2015;30:364–72.
11. Ren Y, Xie Y, Xu Q, Long M, Zheng Y, Li L, et al. University students' fertility awareness and its influencing factors: a systematic review. Reprod Health. 2023;20:85.
12. Beaton DE, Bombardier C, Guillemin F, Ferraz MB. Guidelines for the process of cross-cultural adaptation of self-report measures. Spine. 2000;25:3186.

Chapter 3
CRISPR/Cas9 and the Legitimacy of Nontherapeutic Genetic Intervention

José Tomás Alvarado, Paulo López-Soto, Alex J. Wolf, and Manuel J. Santos

The legitimacy of different forms of genetic intervention has long been the subject of discussion. These discussions have been developed considering science fiction scenarios in which we had the technical capacity to modify the genetic information of an individual at will with full knowledge of the effects of such modification.

The development of the CRISPR/Cas9 (Clustered Regularly Interspaced Short Palindromic Repeats, CRISPR; CRISPR-associated Protein 9, Cas9) technique [1], however, has made all these scenarios seem much closer now and not a theoretically stimulating fiction. In this work, some of the ethical problems that the genetic manipulation of human beings will entail, either through the CRISPR/Cas9 technique or other techniques that can be developed, will be discussed. Many have raised concerns about these procedures, either because a particularly demanding precautionary principle must be kept (cf. [2]) or because the autonomy of the persons on whom this intervention would be made would be affected (cf. [3], with precautions [4]) or because it would be a violation of the attitude of respect and care we owe to nature and the "gift" of life [5]. Of course, some have celebrated CRISPR/Cas9 as one more step to take complete control of our evolution (cf. [6]).

In 1987 was the first discovery of CRISPR locus in *Escherichia coli* (*E. coli*). Since then there has been a development that ended in the CRISPR/Cas9 system, which uses a chimeric single-guide RNA (sgRNA) to direct an endonuclease (Cas9) to specific DNA targets generating double-strand breaks (DSBs). The timeline of CRISPR/Cas9 technique development, from preclinical to clinical phases, is shown in Fig. 3.1, principally in hematological and neuromuscular genetic disorder fields, where it has had the most progress [1, 7].

J. T. Alvarado (✉)
Instituto de Filosofía, Universidad de los Andes, Santiago, Chile
e-mail: jtalvarado@uandes.cl

P. López-Soto · A. J. Wolf · M. J. Santos
Instituto de Bioética, Facultad de Medicina, Universidad Finis Terrae, Providencia, Chile

D. Cerdio-Domínguez, M. E. de los Ríos-Uriarte (eds.), *Current Bioethics*, New Paradigms in Healthcare, https://doi.org/10.1007/978-3-032-25294-4_3

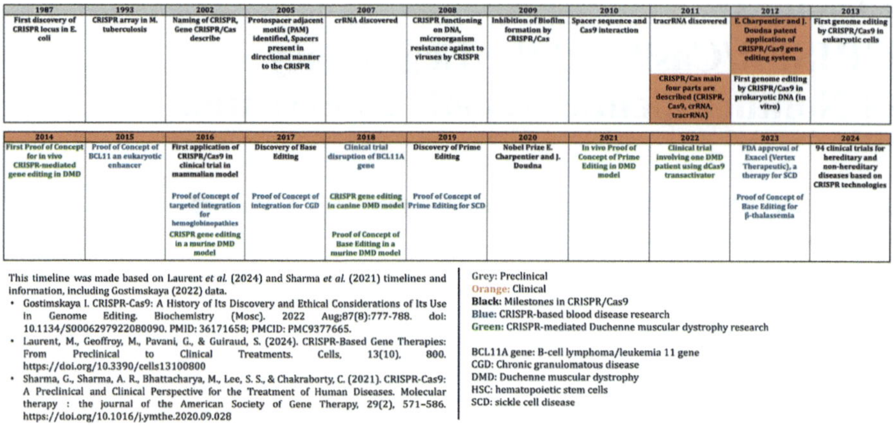

Fig. 3.1 Preclinical and clinical CRISPR/Cas9 timeline

This system is composed of four main parts to make the puzzle, and the final missing piece was discovered in 2011 by Emmanuelle Charpentier's group, a molecule named tracrRNA. Charpentier and Doudna in 2012 applied for a patent of CRISPR/Cas9 gene editing system, a work that assembled in vitro all components in one and was acknowledged by the Nobel Prize in Chemistry in 2020 [7]. It is based on the knowledge that bacteria have regions of DNA that contain sequences called CRISPR (Clustered Regularly Interspaced Short Palindromic Repeats) with genes such as Cas—that encode enzymes called "endonucleases," which cut the DNA—and serving as an immune defense system for these microorganisms, destroying foreign DNAs. Doudna and Charpentier demonstrated that the same molecular system could be used to edit, that is, change the sequence of any DNA in vitro anywhere in the DNA sequence [8, 9]. Thus, the technique has been applied to edit bacteria and animal genes. Yin et al. [10] managed to cure a mouse affected by a lethal genetic disease, and then, Niu et al. [11] applied the technique in monkeys and in human cells [12]. In May 2015, Huang's group applied it to human embryos, publishing an article in *Protein & Cell* Journal after previously having been rejected in *Science* and *Nature* for ethical considerations [13]. In this work, the researchers tried to edit the globin gene, which is part of hemoglobin and is involved in a human genetic disease called "thalassemia." Only in a few cases was it possible to correct the genetic failure. Although the authors noted that the use of CRISPR/Cas9 to edit human embryos "is still premature," this work brought a worldwide discussion, because of its impact and promises. It shows the technical feasibility to change not only defective human genes in embryos or the germline of infants affected by severe genetic disorders—and thus to cure incurable genetic diseases and even eradicate them from the population—but also the technique offers the possibility of adding genes at will and gene editing in children [6, 14, 15].

As a result of all this, a group of Nobel award winners in Medicine and Biology met in the United States in early 2015 and promoted an international moratorium for the application of this technique in humans [14, 16]. Then, the *Red Latinoamericana de Bioética* (RELAB) meeting in Panama also signed this moratorium. In this regard, the scientific world is divided. Some scientists have shown strong opposition to the use of human genome editing. However, others consider that the edition for the cure of diseases is scientifically and ethically justified [17–21]. In April 2016, Kang et al. [22], in order to perform an edition of the genome of human embryos in a specific place, introduced the CCR5Δ32 mutation, related to immunity against HIV, successfully achieving the edition of some embryos. However, modifications of the genome in unintended places were significant, showing low biological safety of the technique. Related to this, in 2018, a Chinese scientist named He Jiankui modified human embryos of fertile couples in which the male parent was HIV positive, for a reproductive purpose. This has a clinical correlate. In 2019 was published a clinical trial where patients with HIV-1 went into remission following CCR5Δ32 hematopoietic stem cell transplantation [23]. With these results that promise therapeutic applications of the technique of genetic editing, England first and then Sweden allowed the investigation of genome editing in human embryos for therapeutic purposes [24, 25]. More recently, in the United States, the use of the germinal genetic edition was approved in limited cases [26].

From a scientific perspective, it is discussed whether the technique is safe enough to be applied to human beings [27–31], until this day, there are ongoing investigations and clinical trials that will address risks and limitations of these system therapeutic applications, which include long-term efficacy, potential genotoxicity, and adverse immune reactions [1, 32].

Until 2024, there were 94 clinical trials for hereditary and nonhereditary diseases using CRISPR techniques [1]. It is the most widely used genome editing tool; it is capable of making gene deletions and insertions, translocations, single-base editing (point mutations are the most common variant associated with human diseases), ARN targeting, gene regulation, and epigenome editing. But, in medical applications, they have potential limitations. To name a few:

- The adeno-associated virus delivery vehicle, which is the most frequently used, has limited packaging capacity of genetic information.
- A remaining major concern is off-target effect. This occurs when Cas9 affects untargeted sites and creates cleavages that may lead to adverse outcomes. Even on-target activation of DNA repair pathways can have adverse consequences on cell viability. For this, there are methods that have been developed for detecting rare mutations and quantifying their potential risks.
- It is reported that Cas protein immunogenicity is another potential obstacle for clinical application [32–34].

In the eventual case of using the CRISPR-Cas9 technique as a treatment for genetic conditions, the classical methodology should be followed, that is, first to test the safety and efficacy in experimental animals or models and then apply in clinical trials in adults and later in children. To eventually apply it to human embryos, in vitro fertilization techniques must be used in which there are additional problems, such as those explained in Ventura-Juncá et al. [35]. Moreover, to establish that an embryo has a condition, they must be subjected to a Preimplantation Genetic Diagnosis (PGD) [36–38]. This technique presents scientific objections, related to their safety, and bioethical repairs [39–43].

There is little doubt that the therapeutic use of CRISPR/Cas9 would be legitimate—if the ethical problems associated with other techniques such as PGD are solved—but concerning nontherapeutic uses, there are moral objections of high weight. It is these reasons that will be discussed in this chapter. In the first section (Sect. 3.1), the reasons why it has been argued that a genetic intervention for nontherapeutic purposes would affect the dignity of the person who has seen their genome manipulated will be presented. These reasons have been answered in different ways, especially by the defenders of transhumanism. From this debate, a position has emerged that has been called "liberal eugenics" that is aware of the risks to human dignity that these techniques have, but, at the same time, admits their legitimacy within specified limits. Many thinkers are inclined to accept this form of eugenics. Then, two types of critical positions regarding liberal eugenics will be examined. One of those is what Jürgen Habermas (Sect. 3.2) has stated, which seems to arise from the same liberal premises that would be recommending liberal eugenics. The problem indicated by Habermas is that liberal eugenics would ultimately be incoherent. The critical examination of Habermas's argument will, however, be negative. Another of these, on the other hand, depends on substantive premises about human good and how human good is related to our nature (Sect. 3.3). A conception of nature as endowed with a normative value imposes additional precautions. This chapter will not discuss the relevance or validity of the central premises of the liberal point of view—which seems to recommend "liberal eugenics"—or the point of view that assigns a normative value to our nature. What will be done to show how these two great models allow for guiding an ethical evaluation of the practices that the CRISPR/Cas9 technique has made possible.

3.1 Liberal Eugenics

A "liberal" conception of public ethics is a conception in which—in general terms—a fair structure of society is a structure that ensures to all persons fundamental equality so that everyone can freely develop their autonomy. In a liberal conception of society, this autonomy is guaranteed by fundamental rights, which include the right to life, physical and psychological integrity, privacy, freedom of thought, conscience and expression, freedom of assembly and association, and so on. Persons endowed with this fundamental equality can participate in the public debate about how society should be organized. It is characteristic of a liberal conception that justice must take precedence over "the good." What is required is to ensure the status of equality so that everyone can pursue their idea of the good life, but a just structure of society does not require that there be a "public" idea of the good on which fundamental equality is grounded. Moreover, the state is required to maintain neutrality about the different comprehensive conceptions of the human good (e.g. [44–48]).

Several have argued that these considerations impose severe restrictions on the practice of genetic engineering. Suppose there are two people, *A* and *B*. *A* decides that it would be good if *B* had less weight, a different form of nose, and another hair color. Without consulting *B*, *A* anesthetizes *B*, takes him to a clandestine pavilion, submits him to cosmetic surgery for which he has liposuction and changes the shape of his nose, and then dyes his hair. Although it is not clear that the changes that *A* has made to *B* are harmful to *B*, it is clear that *A* has acted in a reprehensible way. First, *A* has kidnapped *B* by sedating him and forcibly taking him to a place. He has injured *B*'s physical integrity by making modifications without his consent. The modifications *B* wants to make to his body are something that only *B* should have decided—if it is legitimate that someone could have decided on it, at all. *A* has injured *B*'s autonomy. He has not treated him as a person on an equal footing, able to decide for himself about his own life. It is instructive also to consider a variation of this example. Suppose that, without this being known to *A*, *B* wanted to lose weight, change the shape of his nose and the color of his hair precisely the way *A* has left him after kidnapping him and doing the operations. By chance, it has happened that *A*'s intervention has resulted in something that precisely is what *B* wanted. Even in such a case, what *A* has done has been morally reprehensible, since he has treated *B* not as an equal who can decide on his own life freely and autonomously. The fact that the result has been one that *B* wanted independently does not change the fact that *A* has caused this result by going beyond *B*'s autonomy, denying his dignity as a person on an equal footing. Suppose even that *B* had told *A* that he wanted to lose weight, change his nose shape and the color of his hair. Even in this case, *A*'s action would have been morally reprehensible for the same reasons.

Suppose that *A* and *C* are married and want to have a child. By applying the mechanism of CRISPR/Cas9, they modify the genome of their child *B*, causing him to have blond hair, be 1.95 m tall, and have a specific nose. *B* is then someone with blond hair, 1.95 m tall, thin and with a defined nose. No one has consulted *B* about whether he wanted these traits. They have been imposed by *A* and *C* without considering his opinion. *B* has not been treated as an equal person, able to decide freely and autonomously about his own life. *B*'s dignity seems to have been injured. A nontherapeutic genetic intervention seems to be, then, an injury to the autonomy of people whose genome has been intervened.

This line of reasoning, however, has been answered in several ways [6]:

1. It has been argued that no one has control of its genome. The genetic endowment with which one comes into existence is the product of a "lottery." That lottery is the result of a slow process, full of errors and subsequent high-cost corrections. *B*'s position concerning the genome with which he is born is the same as that of anyone, except that instead of having a genome resulting from chance, he has a genome resulting from conscious planning.
2. If the intervention of *B*'s parents to improve his genetic endowment has been illegitimate, one should also consider illegitimate all interventions that parents make to educate their children. Parents in our cultures decide where their children live, what they eat, how they dress, to what school they go to, whether or

not they have piano or horse-riding lessons, whether or not they can cultivate a friendship, and so on. These are all interventions that have an impact on the future life of a person and have been carried out in the name of parental authority, regardless of the child's will. Parents do not treat their children as equals, then. On the contrary, if the actions taken by parents to educate their children in fulfillment of their parental responsibility are legitimate, so are interventions to improve their genome.

3. Not all genetic interventions should be seen as forms of coercion imposed on the life of a person. There are interventions that, instead of restricting a person's ability to pursue different life plans and different conceptions of human good, enhance their ability to decide what to do with their autonomy freely. A higher intelligence or a taller stature, for example, should not prevent someone from later being a Buddhist or a Catholic, but it would also make it possible for him to be a theoretical physicist if he wanted to, which he could not do without proper intelligence. If someone kidnapped *B* to remove a kidney, this is clearly unacceptable, but it is different if someone gives him a million dollars without consulting him. Giving someone more intelligence is like donating a million dollars and not like removing a kidney.
4. No one has any objection to genetic interventions that aim to cure a disease. The therapeutic use of CRISPR/Cas9 is clearly legitimate. Well, if therapeutic uses are legitimate, interventions should also be admitted to provide a better genome. In effect, we refer to as "therapeutic" any practice that is aimed at correcting, repairing, or restoring an evil for a person, eliminating a source of pain or an impediment that limits the deployment of his autonomy. From a moral point of view, there is no relevant difference between avoiding something bad and actions that positively cause a good. For example, there is no relevant difference between the action that removes an impediment for a person to have a certain IQ and an action that improves the IQ. If so, the legitimacy of therapeutic interventions implies the legitimacy of any intervention that improves the genetic endowment of an individual. On the contrary, if interventions to improve genetic endowment are illegitimate, therapeutic interventions should also be.
5. One must respect the autonomy of existing people able to decide freely about their own life. There is no obligation to respect, however, the autonomy of nonexistent persons, as there are no such persons. No one is obliged to respect the equal dignity of, for example, Don Quixote or Darth Vader because neither don Quixote nor Darth Vader exists. However, the same happens with supposed future generations in which we would be intervening with the present genetic modifications. There are no such people who may be injured.
6. To these considerations must be added that eugenics in the terms envisaged should not have the negative connotations that have eugenic practices of totalitarian regimes of the first half of the last century. It would not be interventions imposed by a state bureaucracy to establish their idea of what is best for people's lives, but rather a practice that would be determined by the parents' free preferences. It would not be eugenics for, for example, removing a certain race, but about eugenics for the good of the children.

7. Furthermore, it seems evident that the application of the principle of beneficence implies that we must use these techniques to achieve the greatest happiness of the greatest number. What we call "human nature" is the product of a slow process, complicated and full of errors. The demands of our life in a rapidly changing environment also require rapid adaptations. What evolution does slowly and imperfectly, we can now do intelligently and quickly. It is not only, then, that it is permissible for us to intervene genetically in future generations, but that we have a moral obligation to do so because we have a moral obligation to ensure the greatest good for the greatest number of people.

Most of these objections can be answered from a liberal perspective, but not all. Several of them are based on a consequentialist moral conception that will not be acceptable. At least it will not be acceptable for author's characteristic of this tradition such as John Rawls (1971), Robert Nozick (1974), or Ronald Dworkin (1978). It is characteristic of a consequentialist position that the only thing relevant to the evaluation of an action is the totality of good or bad consequences that it brings. If there are several courses of action open to a rational agent that action must be carried out that brings with it the greatest amount of goods or, as the case may be, the least amount of evil.[1] It is typical of a consequentialist position that there are no actions that, by themselves, because of the type of action they consist of, are prohibited. Usually, a consequentialist will agree on the legitimacy of respecting the same fundamental rights so important for the liberal tradition, but this is because, in most circumstances, it is the respect for these rights that ensures the best consequences, as these have been determined independently. The same requirement to maximize good consequences, however, makes it mandatory to set aside these fundamental rights in some cases (for an overview [49]). It is not possible to enter here to consider the merits of a consequentialist position. It will be assumed for what follows that certain actions are always morally reprehensible, for example, those in violation of fundamental rights. These fundamental rights are those that make up the status of equality that is required by liberalism for each person.

Objection (1) is based on a consequentialist conception for the moral evaluation of actions. No one is in a position to choose the genome with which it comes into existence, but there is a difference between the intentional act of causing someone to have a certain genome and the fact that someone has a genome. The moral problem arises concerning the genetic intervention that is done intentionally and for which someone comes to be responsible. Nor does anyone choose their death. We will all have to face it. There is a difference, however, between intentionally causing someone's death and someone dying. It is the intentional action that requires moral evaluation. So, just as homicide is an injury to the dignity of a person, a nontherapeutic intervention in a person's genome can be seen as an injury to his dignity.

[1] Assuming that these "calculations" of good or bad consequences make sense. One of the frequent objections to consequentialism is that there is no coherent way of comparing such disparate goods as friendship or aesthetic appreciation of beauty, for example.

From a consequentialist point of view, however, the only thing that matters is the results, not the means that are used for these results.

It is clear that objection (7) also depends on a consequentialist point of view, which allows it to be set aside from a liberal perspective. The obligation to obtain the greatest amount of goods and the least amount of evils regardless of the means used for this result is typically consequentialist. The liberal objection against genetic intervention is precisely highlighting that the action of modifying the genome of another for nontherapeutic purposes is a violation of his dignity. This practice is not being criticized because the genetic endowments that were provoked will have bad consequences. The intentional action of intervening the genome for nontherapeutic purposes is what is reprehensible.

The objection (4) is also dependent on consequentialist premises. There are substantive differences between a therapeutic intervention and an action that simply increases the amount of good consequences.[2] A therapeutic action is an action aimed at restoring health. The concept of "health" is a normative concept that has to do with the proper functioning of a biological organism. It is not the same to make an organism work the way it should work than simply to cause good consequences. If health is identified with good consequences, whatever they may be, obviously the idea of a "therapy" will simply be reduced to nothing. What exists is an obligation of the parents to take care of the health of their children, not the obligation to generate any genetic endowment that can be considered beneficial.

It is also possible to block objections (2), (5), and (6) without too much difficulty from a liberal perspective. As for the objection (2), parents have a responsibility for the education of their children, but it is clear that there is a limit on what is legitimate for them in these matters. The excesses that some parents may commit when they try to supplant the freedom of their children are reprehensible, just as a nontherapeutic genetic intervention is reprehensible. Not anything parents do to educate their children is acceptable (especially, [5]). As for the objection (5), the thesis according to which future people do not exist is very controversial. Many contemporary metaphysicians disagree with this [50]. However, even accepting that future generations do not have the same ontological status as we do, one must admit that we have responsibility for possible people. No one will doubt that future generations are at least possible people. Suppose a hunter observes that something is moving among some bushes. He does not have enough visibility to know if what moves is a deer or a person. According to everything the hunter knows, what moves between the bushes could be a person. Would it be lawful for the hunter to shoot? Obviously not. In any legal system, someone who fired in these circumstances, killing a person would be convicted of homicide. The possibility of a person who may be harmed requires special precautions. Similarly, the possibility of future people whose dignity could be violated requires special precautions. As for the objection (6), it is clear that a scenario in which state

[2] The distinction may in some cases be vague, but this does not prevent the existence of cases that are clearly therapeutic actions and other cases that clearly are not, which is sufficient to make the moral discrimination.

bureaucracies decide on the genetic interventions to be performed would be much worse. This does not, however, mean that nontherapeutic interventions by parents cannot result in a violation of the dignity of the people who suffer from them. The problem is that the intentional action of modifying the genome of another for nontherapeutic purposes is to treat that person not as someone on an equal footing and with full autonomy over himself. This has nothing to do with the question of who is doing the intervention. It would be reprehensible if the state does it. It would be reprehensible if parents do it.

Objection (3) remains, however. Genetic interventions that manage to give the person affected by the intervention a greater range of autonomy so that he can pursue his conception of good afterward do not seem to be an injury to the dignity of the intervened person. It does not seem to be a way of treating another person as an inferior, like someone who is not able to decide freely and autonomously about his own life. This kind of intervention seems to be compatible with the liberal premises about the autonomy of people and neutrality with respect to the conceptions of human good. It is, in effect, a form of eugenics that aims to provide people with greater freedom. It would not be a legitimate form of eugenics that, for example, was aimed at generating people more docile at the orders of the government or with a greater inclination to vote for the Socialist Party or physically weaker people so that they could be submitted with greater ease, but if it is about producing a genetic endowment that expands the sphere of open possibilities in the future, there is no problem with it. This form of "liberal" eugenics has been accepted by many of the most influential liberal theorists [4, 44, 46, 47].

3.2 Genetic Engineering and Our Self-Understanding as Moral Subjects

Jürgen Habermas [3] has argued that the practice of liberal eugenics in the terms set forth brings the risk of undermining the foundations of the same liberal political order. Genetic interventions aimed at providing people with greater autonomy would have the effect of precisely reducing the self-understanding of autonomous subjects whose deliberations sustain the political stability of the liberal order, but also—in the Habermasian perspective—the validity of the normative order of a democratic society. It is not necessary to consider here the peculiar conception of Habermas about how the validity of a norm in a discursive procedure should be justified [51]. The considerations raised about the legitimacy of nontherapeutic genetic interventions are independent of the more general Habermasian theory. Habermas holds:

For the person to feel one with it [with his body], it seems that the body (*Leib*) has to be experienced as something natural, such as the continuation of the organic, self-regenerating life, from which the person was born. We live our freedom as referring to something naturally unavailable [3].

Only the reference to this difference between nature and culture, between unavailable beginnings and historically modeled practices, allows the agent performative self-attributions without which he could not understand himself as the initiator of his actions and claims [3].

The liberal conception that has been mentioned above understands that justice demands that each person is assured a fundamental equality so that each one can autonomously display his or her conception of the human good. What Habermas adds to this traditional liberal perspective is that: (i) It is required that rational subjects be able to understand themselves as such, that is, as subjects on an equal footing who freely decide on their own lives. Then, (ii) that for this self-understanding, the difference between the natural and the cultural is required, between what is unavailable and given, and what is a product of our free action. Nontherapeutic genetic interventions would be erasing the difference between the natural and the cultural—because everything would be an artifact of our strategic planning—and, thereby, erasing the fundamentals of free action.

These two premises require a closer critical examination.

- A liberal society is organized on a status of fundamental equality granted to each person. People who can freely deploy their life plan are also in a position to deliberate about public issues. They must be subjects, then, who freely decide according to reasons, which can give reasons for action and can criticize the reasons offered by others. A condition of possibility for public deliberation is that all participants can recognize the other participants as rational subjects. The communicative practice of dialogue with someone presupposes that the interlocutor is someone who can understand our reasons and is someone whose reasons can be understood by us. For example, it would be bizarre to find someone giving a lengthy explanation to his cat about the economic benefits of free trade. A person who takes his cat seriously as an interlocutor on an equal footing, who sincerely hopes that the reasons he can explain to his cat will have an impact on the cat's decisions, is someone we would say has lost his mind. We do not engage in conversations with a cat because they make no sense. It can be appreciated, then, that the practice of engaging in a dialogue presupposes that one sees in the interlocutor someone capable of acting freely for reasons, but—at the same time—one must presuppose that the interlocutor can see one as a subject able to decide freely for reasons. The interlocutor can be moved to accept the reasons one can give him because he must, at least, be able to see us as rational subjects who deliberate on reasons, able to communicate and respond to them. One cannot presuppose that an interlocutor is going to take one as a rational subject who acts for reasons without self-understanding oneself also as a rational subject who acts for reasons.
- The second thesis proposed by Habermas is that one cannot understand oneself as a free subject without presupposing the distinction between the natural and the cultural. The "natural" must be understood as everything that is given to us, and

that is not available to be transformed by our free activity.[3] The "cultural," on the other hand, is everything that has resulted from the free actions of rational subjects. The genome with which one comes into existence has always been part of the "natural": a domain not available for our intervention and that defines—at least in part—the scope of what one can decide freely to do. It would make no sense, for example, that one was deliberating whether to swim like a fish or not because we do not have gills. It is not part of our genome, and nothing one can do about it will change this. The experience of a fish in the water is not an experience that one might have. The deliberation about what actions one is going to perform makes sense within the limits of actions that can be performed. For this reason, it is that the "natural" establishes a limit of that over which makes no sense to deliberate, which is, at the same time, what opens the space of that which can be deliberated.

It has been common since Modernity to think of the human as the deployment of complete autonomy. It has been usually assumed that the main characteristic of a human being is freely to be what he wants to be. Human freedom, however, is not omnipotence. It has limits that are inevitable in the case of a finite subject. These limits not only fulfill a "negative" function but also fulfill a "positive" function. It cannot be otherwise when it comes to the nature of finite freedom.[4] Our freedom is a power or capacity that must be grounded on certain characteristics or traits that, together with opening the space of possibilities that can be accessed with such power or capacity, imply its limits. Being human implies being neither a fish, nor a bird, nor a dinosaur, nor a supernova. The same properties that open for us a space of possibilities for our freedom close other possibilities incompatible with such properties. When it comes, then, to finite freedom—like ours—if there were nothing given "natural," unavailable, there would be no freedom for us. Our self-understanding, as finite subjects, follows these structural features of our finite freedom. We do not deliberate on everything. Not everything is open to our autonomous action. What is open to our action is so because there are things that are not and of which we are aware that they are not.

The premises (i) and (ii) taken together imply that by erasing the distinction between the "natural" and the "cultural," the liberal order that requires rational and free subjects that self-understand as such is also undermined. However, this seems clearly excessive. The genetic intervention does not entirely erase the difference between "natural" things that we cannot dispose of and "cultural" things that we do. What it does is simply modify the limit between these domains. Something that was previously taken as something given concerning which nothing can be done becomes something that we can manipulate. This, in itself, does not destroy our finite

[3] This is a fairly simplified way of indicating what 'natural' implies in this context. The concept of 'nature' has had a very complex history. These schematic indications should be related to the observations that will be made below.

[4] Moreover, talk here of the "nature of freedom" is not an oxymoron. Our freedom is one of the natural things that we cannot dispose of.

freedom. The practice of genetic engineering via CRISPR/Cas9 will not make, for example, the laws of physics cease to be as they are. Genetic interventions may cause us to have a different endowment than our ancestors, but surely it will not allow our children to be, for example, dinosaurs or bacteria. Secondly, this process of "modifying the boundaries between the natural and the cultural" is something that human beings have been doing for thousands of years. It is certainly something that has been done when animal grazing and agriculture have developed. It is something that has been done by the invention of the arch in architecture, by developing antibiotics and our cell phones. If the practice of genetic intervention in human beings is reprehensible for "eliminating the distinction between the natural and the artificial" and, thereby, for "undermining our self-understanding as free subjects," we should also condemn for the same reasons our ancestors who for the first time began to domesticate animals and plants, instead of hunting them and collecting them in the wild.

It seems, then, that Habermas's critique of liberal eugenics is assuming that genetic intervention will completely eliminate the domain of the "natural," which is clearly not the case. Only the complete elimination of the "natural" would bring the catastrophic result that we would not recognize ourselves as finite free subjects. But genetic intervention is far from such a thing. The CRISPR/Cas9 technique makes us more powerful but not omnipotent.

3.3 Respect for *Our* Nature

Habermas's criticism is presented as an "internal" critique of the liberal point of view about nontherapeutic genetic interventions, that is, supposedly the so-called "liberal eugenics" is ultimately in conflict with the premises that support it. There is, however, a critical perspective that depends on different premises. It is interesting to note that several authors who have not expressly accepted such premises seem to share the discomfort regarding nontherapeutic genetic interventions that follow from them. Perhaps this explains the attitude of Habermas regarding liberal eugenics. One of these authors is Michael Sandel, who concerning the legitimacy of these interventions points out:

The problem with eugenics and genetic engineering is that they represent the unilateral triumph of the will over the gift, of the dominion over reverence, of manipulating over contemplating [5].[5]

[5] Sandel also points out: "To suppose that our talents and gifts are completely our work is to misunderstand our position in creation, to confuse our role with that of God. But religion is not the only source of reasons to worry. Moral concern can also be described in secular terms. If the genetic revolution erodes our appreciation of the gifted character of human powers and achievements, it will transform three crucial features of our moral landscape –humility, responsibility and solidarity." (Sandel, 2007, 85–86). An eventual genetic revolution would transform our "moral landscape" according to Sandel, because our capabilities should be attributed to human planning

A genetic revolution not only would be a triumph of our manipulative attitude over the attitude of humble gratitude for a "gift" received—as says Sandel—but it would be the climax of such manipulative attitude. In the same way, not only it would be a further modification of the boundaries between nature and culture, but a radical change of what "culture" means and, therefore, would be a severe threat to our self-understanding as free subjects. Each technical advance has been a triumph of our will against what is given to us. Each technical advance has been an extension of the "culture" over the merely "natural." The deployment of human freedom is not, in itself, reprehensible. Those deployments have not been reprehensible, as the development of agriculture, the construction of cathedrals, the development of steam engines or the personal computer. When it comes to nontherapeutic genetic engineering, however, there is a crucial difference from all the indicated cases of technical progress. This difference is what makes it the paroxysm of the impulse to affirm our will to power. This difference is what makes it seem like a claim to supplant God and is what makes genetic engineering morally reprehensible. The substantive difference between intervention in our genome and any other technical development is that it aims at the transformation of *our nature*.

One of the central contrasts between this alternative point of view and the liberal point of view has to do with the concept of "nature" that is operating in the moral evaluation of genetic engineering. In Classical Antiquity, *physis* or *natura* designates an intrinsic principle of operations for a substance that is teleologically directed. In Modernity, however, the "natural" has tended to designate a teleologically blind mechanism that can be manipulated for the purposes that a rational and free subject wants to impose (for a history of the contrast [52]). Western philosophical tradition has sustained that what we call "human nature" is not merely a set of biological facts, but something endowed with a value for being directed toward certain ends that give meaning to the deployment of natural dynamisms. Our corporeality is given from the beginning with certain inclinations that are guides for, for example, our assessment of life and the sexual union. Our inclinations are extended from the beginning to the appreciation of truth, beauty, and friendship. Basic human goods seem to be "calibrated" with the inclinations that make us appreciate them [53–56]. A defense of this theoretical perspective is not intended here—as well as a defense of the liberal point of view has not been made either. What is interesting is to highlight how this perspective deeply rooted in the cultural tradition of the West imposes reasons for resistance against the practice of nontherapeutic genetic intervention. The hypothesis of an intervention on human nature is morally problematic

in the future. Therefore, any lack in these capacities will be something for which someone will be responsible. Any deficiency or ugliness in our corporeality will be the fault of someone who has not done the planning that he has been able to do. A child with Down syndrome is not something that will claim our solidarity, but rather an imprudence or carelessness for which someone should be held responsible. Our lives will appear not as a gift, but as an artifact product of the manipulation carried out by some planner. The responsibility of these planners will be extended to each feature of our genomes and, in the same measure, will decrease the inclination of the best gifted to compensate for the situation of the least gifted. After all, why would the most genetically gifted have to take responsibility for the mistakes of others?

because our nature is also a normative guide. The claim to reconfigure our own nature is, then, the claim to become lords of good and evil, since the normative standards that should be the guide for future generations will be those imposed by certain planners—that such planners are parents, grandparents, or state bureaucrats is indifferent here.

There are at least two serious problems with genetic engineering from this perspective. In the first place, it manifests an enormous arrogance to suppose that we can plan the traits of future human beings. We do not know enough about, for example, epigenetic factors and how these factors interact with the genome for the development of an organism. Even from a purely naturalistic perspective, it is excessive to assume that a human planner will be able to improve what has been achieved with millions of years of evolution. No one really knows whether to "endow" a person with certain traits will imply or not "expand their autonomy." No one really knows if making people smarter or better equipped to play the piano is not going to be ways of narrowing the possibilities of human fulfillment of those future people. To think that we can control all the consequences at stake is, again, arrogance. Moreover, it is arrogance that can cause serious harm. For the same reasons, we do not know how we would "isolate" a nontherapeutic genetic intervention of, for example, intelligence without affecting someone's sense of friendship or beauty or truth. We do not know how the interventions in our corporeity can affect the delicate calibration by which our inclinations are moral guidelines for judgments about the good and the bad.

There is a second additional problem, from this perspective. A transformation in our fundamental moral inclinations would not only have an impact on future generations in what has to do with the greater or lesser clarity about certain ethical judgments. There would be a problem in the style of what Habermas argued. Our self-understanding as rational and free subjects would be obscured. This problem, however, would not arise because genetic engineering would eliminate the distinction between "natural" and "cultural," as Habermas seems to have suggested, but because the modification of our own nature, the pretense of converting our nature into an artifact, would affect the conditions of possibility of our self-understanding. Our self-understanding as rational and free subjects presupposes our nature as a normative guide. It is the dislocation of human nature that has dangerous effects. It is typical of finite freedom that not everything has to be subject to free decision. Not only because there are things that our action cannot change, but because deliberation is possible within a framework of already given volitions. Our rational deliberation has to do with the consideration of means and ways in which we can participate in basic human values, but it is not about the goodness of those values and our inclination to want to participate in them. We need certain volitions already given that are not subject to deliberation so that deliberation makes sense. This thesis has been expressed in several ways. It has been said, for example, that every human being wishes to be happy and desires good. The tendency to the good that is not subject to deliberation, but is a condition of possibility for specific free deliberations and decisions, has been termed as *voluntas ut natura* [57] as opposed to the *voluntas ut ratio*.[6] In any deliberation, an end must be presupposed that is not in question, and that functions as a horizon for the rational consideration of the different alternatives that are open to the will. If one is deliberating whether to make a trip by bus or plane, it is because one has already decided to make a trip. Of course, one could also deliberate about whether or not to make the trip. Free decisions already made function as a horizon in which other subsequent decisions will make sense. The point is that not everything is subject to deliberation. Nobody deliberates whether or not he wants to have a happy or full human life. No one deliberates whether he will seek the good in his life or not. This fundamental volitional tendency is given to us as part of our nature and is, as explained, a fundamental guide to our moral orientation in the world. We do not know if genetic intervention, supposedly aimed at expanding the autonomy of future generations, is not going to dislocate this sense of moral orientation.

As explained above, it is a condition for the possibility of a rational dialogue that we can recognize other people as persons capable of giving and responding to rea sons. It is a condition of possibility that we can also recognize ourselves as such people who respond to reasons. It has been emphasized by many thinkers that the practice of interpretation by which we can "make sense" of others and what they say or express requires a delicate process of "triangulation." We get to have adequate evidence that someone has uttered a statement with a certain meaning because we must assume that he can know the salient features of his environment as we can know them and because we must assume that he has interests and inclinations similar to ours [58]. Interpretation practices presuppose that we can contemplate our interlocutors as—in some way—participants of the same community in the truth and the good. The dislocation of the moral guide that constitutes our fundamental natural inclinations, then, could generate the estrangement of those who do not share them—or of those of whom we cannot assume that they share them. This would be a difficulty to make sense of other people as subjects who respond to reasons and would, correspondingly, be a difficulty for our self-recognition as subjects who respond to reasons.

3.4 Conclusions

In this work, two main lines of moral questioning for genetic engineering practices for nontherapeutic purposes have been presented, which have acquired a new prominence with the recent development of the CRISPR/Cas9 technique. The first line of questioning comes from liberal premises according to which every person should be granted equal status for the autonomous development of any life plan they want to

[6] That is, the will can be taken as an inclination or a "natural" will (*ut natura*), which is not the object of free decision, but also as the inclination that arises after rational deliberation (*ut ratio*) about the most appropriate ways to participate in basic goods or about appropriate means to reach such participation.

develop. In this line of moral evaluation, the problem with CRISPR/Cas9 is the injury to the autonomy of the people that could result from it. It is compatible with this perspective, however, a genetic engineering practice that does not close the future possibilities for a person to pursue any idea of good. Habermas has argued that this form of "liberal eugenics" is, however, incompatible with liberal premises, since it would affect the self-understanding that rational agents would have as autonomous subjects. It has been shown, however, that Habermas' argument is insufficient to justify this conclusion.

The second line of moral evaluation of CRISPR/Cas9 is much more severe. This second line depends on premises very different from the liberal premises indicated. It is a normative conception of human nature according to which a certain moral guide for an accomplished life or a good life is inscribed in it. From this perspective, the claim to modify our own nature is seriously reprehensible, since it is about manipulating the moral guidance that is inscribed in our natural inclinations. It is curious to note that in this conception there is a certain risk that genetic engineering—carried out through CRISPR/Cas9 or another technique—causes what Habermas feared: the darkening of our self-understanding as rational and free subjects.[7]

References

1. Laurent M, Geoffroy M, Pavani G, Guiraud S. CRISPR-based gene therapies: from preclinical to clinical treatments. Cells. 2024;13:800.
2. Peters T. CRISPR, the precautionary principle, and bioethics. Theology Sci. 2015;13:267–70.
3. Habermas J. El futuro de la naturaleza humana. ¿Hacia una eugenesia liberal? Barcelona: Paidós; 2002.
4. Buchanan A, Brock DW, Daniels N, Wikler D. From chance to choice: genetics and justice. Cambridge: Cambridge University Press; 2000. https://doi.org/10.1017/CBO9780511806940.
5. Sandel MJ. The case against perfection: ethics in the age of genetic engineering. Cambridge, MA: Harvard University Press; 2007.
6. Harris J. Germline manipulation and our future worlds. Am J Bioeth. 2015;15:30–4.
7. Gostimskaya I. CRISPR–Cas9: a history of its discovery and ethical considerations of its use in genome editing. Biochemistry. 2022;87:777–88.
8. Chylinski K, Le Rhun A, Charpentier E. The tracrRNA and Cas9 families of type II CRISPR-Cas immunity systems. RNA Biol. 2013;10:726–37.
9. Niewoehner O, Jinek M, Doudna JA. Evolution of CRISPR RNA recognition and processing by Cas6 endonucleases. Nucleic Acids Res. 2014;42:1341–53.
10. Yin H, Xue W, Chen S, et al. Genome editing with Cas9 in adult mice corrects a disease mutation and phenotype. Nat Biotechnol. 2014;32:551–3.
11. Niu J, Zhang B, Chen H. Applications of TALENs and CRISPR/Cas9 in human cells and their potentials for gene therapy. Mol Biotechnol. 2014;56:681–8.

[7]This work has been written in execution of Project VRI 6063/2016, Pontificia Universidad Católica de Chile. We appreciate the comments and suggestions to the previous versions made Luca Valera and Miguel González.

12. Niu Y, Shen B, Cui Y, et al. Generation of gene-modified cynomolgus monkey via Cas9/RNA-mediated gene targeting in one-cell embryos. Cell. 2014;156:836–43.
13. Liang P, Xu Y, Zhang X, et al. CRISPR/Cas9-mediated gene editing in human tripronuclear zygotes. Protein Cell. 2015;6:363–72.
14. Blasimme A, Anegon I, Concordet JP, et al. Genome editing and dialogic responsibility: "What's in a name?". Am J Bioeth. 2015;15:54–7.
15. Werner-Felmayer G, Shalev C. Human germline modification—a missing link. Am J Bioeth. 2015;15:49–51.
16. Baltimore D, Berg P, Botchan M, et al. A prudent path forward for genomic engineering and germline gene modification. Science. 2015;348:36–8. https://doi.org/10.1126/science.aab1028.
17. Cyranoski D. Ethics of embryo editing divides scientists. Nature. 2015;519:272.
18. Germline editing: time for discussion. Nat Med. 2015;21:295.
19. Kaiser J, Normile D. Embryo engineering study splits scientific community. Science. 2015;348:486–7.
20. Scott C. Treading the line between sensational and groundbreaking science. Am J Bioeth. 2015;15:1–2.
21. Vogel G. Embryo engineering alarm. Science. 2015;347:1301.
22. Kang X, He W, Huang Y, et al. Introducing precise genetic modifications into human 3PN embryos by CRISPR/Cas-mediated genome editing. J Assist Reprod Genet. 2016;33:581–8.
23. Salas SP. Edición genética humana: análisis de algunos desafíos éticos. Rev Med Clin Condes. 2022;33:140–7.
24. Callaway E. UK scientists gain licence to edit genes in human embryos. Nature. 2016;530:18.
25. Callaway E. Gene-editing research in human embryos gains momentum. Nature. 2016;532:289–90.
26. Kaiser J. A yellow light for embryo editing. Science. 2017;355:675.
27. LaBarbera AR. Proceedings of the international summit on human gene editing: a global discussion—Washington, D.C., December 1–3, 2015. J Assist Reprod Genet. 2016;33:1123–7.
28. Lundberg AS, Novak R. CRISPR-Cas gene editing to cure serious diseases: treat the patient, not the germ line. Am J Bioeth. 2015;15:38–40.
29. Mertes H, Pennings G. Modification of the embryo's genome: more useful in research than in the clinic. Am J Bioeth. 2015;15:52–3.
30. Hildt E. Human germline interventions—think first. Front Genet. 2016;7:81. https://doi.org/10.3389/fgene.2016.00081.
31. Ishii T. Reproductive medicine involving genome editing: clinical uncertainties and embryological needs. Reprod Biomed Online. 2017;34:27–31.
32. Sharma G, Sharma AR, Bhattacharya M, et al. CRISPR-Cas9: a preclinical and clinical perspective for the treatment of human diseases. Mol Ther. 2021;29:571–86.
33. Pickar-Oliver A, Gersbach CA. The next generation of CRISPR–Cas technologies and applications. Nat Rev Mol Cell Biol. 2019;20:490–507.
34. Guo C, Ma X, Gao F, Guo Y. Off-target effects in CRISPR/Cas9 gene editing. Front Bioeng Biotechnol. 2023;11:1143157.
35. Ventura-Juncá P, Irarrázaval I, Rolle AJ, et al. In vitro fertilization in mammals: epigenetic and developmental alterations. Biol Res. 2015;48:68.
36. Collins SC. Preimplantation genetic diagnosis: technical advances and expanding applications. Curr Opin Obstet Gynecol. 2013;25:201–6.
37. Chen CK, Yu HT, Soong YK, Lee CL. New perspectives on preimplantation genetic diagnosis and preimplantation genetic screening. Taiwan J Obstet Gynecol. 2014;53:146–50.
38. Yan L, Wei Y, Huang J, et al. Advances in preimplantation genetic diagnosis/screening. Sci China Life Sci. 2014;57:665–71.
39. Rethoré MO. Prenatal and preimplantation diagnosis from the parents' viewpoint. In: Sgreccia E, Laffitte J, editors. The human embryo before implantation: scientific aspects and bioethical considerations. Vatican City: Libreria Editrice Vaticana; 2007. p. 167–76.

40. Ventura-Juncá P, Santos MJ. The beginning of life of a new human being from the scientific biological perspective and its bioethical implications. Biol Res. 2011;44:201–7.
41. Hens K, Dondorp W, Handyside AH, et al. Dynamics and ethics of comprehensive preimplantation genetic testing: a review of the challenges. Hum Reprod Update. 2013;19:366–75.
42. Whetstine LM. Ethical challenges in assisted reproduction: the place of preimplantation genetic diagnosis in a just society. J Child Neurol. 2015;30:547–50.
43. Damian BB, Bonetti TCS, Horovitz DDG. Practices and ethical concerns regarding preimplantation diagnosis: who regulates preimplantation genetic diagnosis in Brazil? Braz J Med Biol Res. 2015;48:25–33.
44. Rawls J. A theory of justice. Cambridge, MA: Harvard University Press; 1990.
45. Rawls J. Liberalismo político. Barcelona: Criticism; 1996.
46. Nozick R. Anarchy, state, and utopia. Oxford: Blackwell; 1974.
47. Dworkin R. Sovereign virtue: the theory and practice of equality. Cambridge, MA: Harvard University Press; 2000.
48. Sandel MJ. El liberalismo y los límites de la justicia. Barcelona: Gedisa; 2000.
49. Scheffler S, editor. Consequentialism and its critics. Oxford: Oxford University Press; 1988.
50. Curtis B. A critical introduction to the metaphysics of time. London: Bloomsbury Academic; 2016.
51. Habermas J. Facticidad y validez, Sobre el derecho y el estado democrático de derecho en términos de la teoría del discurso. Madrid: Trotta; 1998.
52. Spaemann R. Lo natural y lo racional. Santiago: IES; 2011.
53. Finnis J. Natural law and natural rights. 2nd ed. Oxford: Oxford University Press; 2011.
54. Gómez-Lobo A. Bienes humanos. Ética de la ley natural. Santiago: Mediterráneo; 2006.
55. Lee P, George R. Body-self dualism in contemporary ethics and politics. Cambridge: Cambridge University Press; 2008.
56. García-Huidobro J. Razón práctica y ley natural. El iusnaturalismo de Tomás de Aquino. Valparaíso: Edeval; 1993.
57. de Aquino T. Summa theologiae: cum text ex recensione leonina. Roma: Marietti; 1956.
58. Davidson D. Radical interpretation. Dialectica. 1973;27:313–28.

Part II
Clinical Bioethics

Chapter 4
Current Structures, Practice, and Impact of Clinical Ethics Consultation in Healthcare

Becket Gremmels

4.1 Introduction

Clinical ethics consultation is emerging as an allied health field in the United States and across the world. A review of prominent options for structuring clinical ethics within a health system, developments in the practice of clinical ethics, and its impact on the dignity of patients and clinicians as well as hospital operations would help the field identify best practices and improve as it grows. The following summary of the current state of the field reflects my own view based on my experience and discussion with colleagues; it is not meant to be exhaustive or comprehensive of all possible methods to structure and practice clinical ethics. Even still, it should prove helpful to those who are new to the field, to any facility that is revising their structure or practice, or to anyone trying to describe or quantify the impact of clinical ethics in the clinical setting.

B. Gremmels, PhD (✉)
CommonSpirit Health, Chicago, IL, USA
e-mail: becket.gremmels@commonspirit.org

D. Cerdio-Domínguez, M. E. de los Ríos-Uriarte (eds.), *Current Bioethics*, New Paradigms in Healthcare, https://doi.org/10.1007/978-3-032-25294-4_4

4.2 Structures of Clinical Ethics Programs

The structure of clinical ethics programs varies widely across hospitals and health systems.[1] The differences are so significant that it is difficult if not impossible to make meaningful comparisons between programs. The field greatly needs more research on what structures are currently in use, how widespread they are, and what features (if any) are common across programs. An evaluation of the effectiveness of these structures would also be invaluable, but a descriptive analysis is needed first. Some of this descriptive work may emerge from the Clinical Ethics Consultation Benchmarking Collaborative (CECBC), which collects data on clinical ethics consults and ethics consultation services at hospitals across the United States [1]. Yet, a brief summary of some of the more common structures will have to suffice until such work is completed. To be clear, this description is based on my own experience, supplemented by the literature, and is not based on data or surveys of the field because such data do not exist. Consequently, it includes broad generalizations that may lack nuance or not fully describe all programs. That being said, I will delineate structures according to two categories: employment model and scope.

There are an estimated 27,000 people doing clinical ethics work in the United States [2]. Among these, four employment models are common for ethics programs.[2] First is the Professional model. Many ethics programs employ professionally trained ethicists. In these programs, all ethics work is done by these professional ethicists. Since there is no requirement or standard training to become an ethicist, what constitutes a "professional ethicist" varies. Training may include academic degrees in bioethics (a doctorate or master's program), experiential training like a fellowship or internship, or a combination of both. Some professional ethicists are clinicians who shift into ethics, while others have nonclinical backgrounds like theology, philosophy, or sociology. Two efforts to standardize the path to becoming a professional are the Healthcare Ethics Consultation certification (HEC-C) and the Council on Program Accreditation for Clinical Ethics Training (COPACET) [3, 4]. However, both of these are still in the early stages; neither is a requirement or even a standard approach. There are at least two variations in the reporting structure of the Professional model; some professional ethicists are employed directly by the local facility, while others are employed by the health system or region and are deployed locally. Some Professional models employ ethicists full time; others employ them half time or part time. Of those employed half time or part time, they often spend the rest of their time on academic teaching, on clinical service in another profession if they are also a practicing clinician, or they may have another job

[1] I use ethics program to mean the overall structure providing any and all ethics activities within an organization. I use ethics consultation service to mean the service that provides ethics consultation underneath the broader umbrella of an ethics program.

[2] A potential fifth "model" might be to have no one doing ethics work, not even unpaid volunteers. Unfortunately this is the case at many hospitals. I do not include that "model" here since it would describe the lack of structure rather than the structure of an actual ethics program.

outside the institution. It is unclear how many professional ethicists there are, but in my experience, the numbers have increased over the past decade.

Second is the Volunteer model. In this model, the people doing ethics work are not paid for their time devoted to that work. Usually, these people are employed by the hospital in another role (chaplain, social worker, physician, etc.) and spend some of their time doing ethics work. This is not a part of their official job description. It is something they do simply because they feel called to it or were asked to. Some Volunteer models rely on people who are not employees, such as professors at a local university or a community member who has experience in ethics. Either way, these people volunteer their time to perform and support the ethics work at the facility. Most of them have no professional training in ethics, like a degree or fellowship. Many have skills that they have acquired through years of experience. Some may have no training or skill sets related to ethics work. It is unclear how many facilities use the Volunteer model or how many people are doing ethics work under this model.

Third is the Dedicated Time model. Similar to the Volunteer model, the Dedicated Time model involves employees of the facility doing ethics work, but they are paid for that work. Ethics is part of their official job description in some fashion, and they have some amount of time dedicated to that work that could range from a few hours a month to a few hours a day. While most of them may not have a degree or formal training in ethics like a fellowship or internship, they may have completed a certificate program in ethics or attended some workshops. Regardless of their background, they have some formal designation to do ethics work and time in their job that is protected for and dedicated to that work.

Fourth is the Hybrid model. This is a combination of two or more of the other three models. Hybrid ethics programs may have some professional ethicists and have some people who volunteer their time or some people with time dedicated to ethics or some combination thereof. This may involve a professional ethicist at a regional level who is available for more complex consult or non-consult activities, or it could involve a professional ethicist who performs most ethics work and has a few people under one of the other models to serve as backup or take call occasionally. Some Hybrid models do not have any professional ethicists and are instead comprised of a combination of volunteers and those with dedicated time to ethics.

All four of these models can also vary in their scope: single facility, regional, system wide, and a combination of the three. First, some ethics programs cover a single facility. These programs primarily serve one facility. Second, ethics programs may cover a region with multiple facilities. These vary in size from a few square miles to hospitals over 1,500 miles apart or more. Third, a program may be at the system level of a health system. Some systems may only have one person at the system level who is responsible for the whole system. There may also be multiple people at the system level who divide up responsibility by geography, service line, type of ethics work, or some other criteria. Lastly, some health systems have a combination of these scopes with people doing ethics work at local, regional, and system levels. There are some efforts to describe staffing standards to determine when a single facility should have a dedicated ethicist, or how many facilities a regional ethicist can or should cover, but there are no accepted standards at this time [5]. Even still, it appears that bedside clinicians prefer an embedded ethics model, with one or more dedicated ethicists at larger facilities. An embedded ethicist is one who attends clinical rounds, is present on the units and department meetings, and is available for in-person family meetings. As one physician stated, "be in the trenches with us. Only then will we appreciate you and use you as a service. Otherwise, we will think you're speaking to us 'on-high' and will basically ignore everything you suggest" [6]. This scope facilitates the humanization of medicine by respecting the dignity of clinicians through moral support.

These four categories of scope can be subdivided into three forms of focus: Single Focus (one or more people doing ethics work at each facility), Shared Focus (one or more people share coverage for more than one facility), and Hub and Spoke (primary focus at one facility, usually a large referral center, with availability to cover other care sites as issues arise). Any of the four categories of scope and any of the three forms of focus can be combined with any of the four employment models, leading to at least 48 potential structures. More research is necessary to delineate other possible structures, identify how often each one is used, and assess their effectiveness in different environments.

4.3 The Practice of Clinical Ethics

The practice of clinical ethics varies almost as much as the structures of ethics programs. There is no standard list of what counts as the practice of clinical ethics, often called "ethics activities," but some actions are certainly widespread. The three traditional activities include clinical ethics consultation, ethics education, and policy review. A comprehensive overview of these activities would be too long for one chapter, so I will instead mention some best practices. First, ethics consultation has historically been reactive and still often is so today. The ethics consultant waits for someone to request a consult before they get involved in a patient's care. Conversely, a proactive consultation approach attempts to identify the need for a consult before a request is made. This allows the ethics consultant to either provide an ethics intervention before a situation worsens or provide the patient, surrogate, or care team with tools, language, or resources to resolve the issue on their own [7].

Second, ethics education for clinicians is often haphazard, based on recent articles or cases in the news. A data-driven approach analyzes trends in recent ethics consultations to identify education needs. For example, if one unit has had an increase in consults related to identifying surrogate decision-makers, the ethics program should offer an education session on the applicable law and policy. Third, the ethics program should be a required reviewer for ethics-related policies, but the policy should be owned by the relevant clinical department. For example, any changes to a policy on Do Not Resuscitate (DNR) orders should be reviewed and approved by the ethics program, but a clinical department like critical care or medical staff services should own the policy. The ethics program should weave values and norms into policies as relevant, as well as references to ethics-related statements by professional organizations or other prominent literature.

Two newer core functions include community outreach and process improvement. Community outreach might involve connecting with community members and leaders to provide ethics education or work on projects like advance care planning or working with local courts to educate court-appointed guardians on ethical considerations when making medical decisions for a patient. Good education results in decisions that respect the patient as a fellow human being rather than a decision made to mitigate risk. Process improvement involves changing the workflow of clinicians to reflect ethics best practices, articles, or consensus statements in order to prevent ethical issues from arising to begin with or providing clinicians with the tools to resolve ethical concerns on their own. For example, it is common practice to temporarily revoke DNR orders before surgery. An ethics process improvement might involve changing informed consent forms and pre-op checklists to make sure that the surgeon and patient (or surrogate) are all in agreement about intraoperative code status before surgery begins, in accordance with the statement by the American Society of Anesthesiologists [8]. Other activities include organizational ethics consultation, academic teaching (which is distinct from education for clinicians or employees), moral distress consults, facilitating leadership discernment, research and scholarship, managing and developing the ethics program, and professional development [9].

Data collection and analysis is another ethics activity of growing importance. Ethics programs are increasingly using data to help determine priorities and strategy for ethics activities and initiatives, which requires tracking data. Most data tracking currently centers on clinical ethics consultation. The patient's medical record is a clear source of data, and most ethics consultation services also track internal data that are not in the medical record. Data also allow for comparison of different facilities served by the same ethics consultation service and between different ethics consultation services. However, the services must track similar datasets for a meaningful comparison. There are a few efforts to create a minimum standard set of data elements to track, but there is no single standard as of yet [1, 10]. The more ethics programs that track these minimum data points, the easier it will be to evaluate and compare practices. Some ethics programs also track data related to education, organizational ethics consults, discernment, and other activities. Analysis of these activities is still in its infancy and is an area worth more attention in the literature. A focus on data might seem to dehumanize a sacred, personal experience like clinical ethics. After all, the calculating analysis of statistics contrasts sharply with the warmth of an ethics consultant who listens and responds to your concerns. However, data provide a necessary foundation for ethics programs to expand their efforts to defend the dignity of patients and clinicians and steward ethics resources wisely.

Despite the lack of standards in data tracking, some best practices are emerging as the field of clinical ethics continues to professionalize. Some examples include Striving for Excellence in Ethics from the Catholic Health Association (CHA), which focuses on the practice of an entire ethics program, and Core Competencies for Healthcare Ethics Consultation from the American Society for Bioethics and

Humanities (ASBH), which focuses more on the practice of individual ethicists and also includes some elements of program evaluation. Other tools like Assessing Clinical Ethics Consultation Skills (ACES) can help evaluate individual clinical ethics consults, and competency milestones can be used to assess the development of an ethics consultant over time [11, 12].

While these documents lay some groundwork for professionalization, there clearly is no widespread consensus on practice standards as of yet. Many areas have wide variation. For example, some clinical ethicists never meet with the patient, surrogate decision-maker, or the patient's family on principle. Others make a point to do so in every consult. Some meet with the patient or surrogate separately, without the care team present while others only meet with the patient or surrogate in conjunction with the care team. There is not even a standard definition of what constitutes a clinical ethics consult, much less a standard process for counting consults, or a taxonomy of types of consults. Even the model for consultation itself varies: some ethics consultations services prefer the individual consultant model, in which one person performs the consult, others rely on a small team of consultants, and some gather an entire ethics committee for each consult. While all three models still occur, only 16% of hospitals generally use the full ethics committee model for most consults [2].

One recent practice that is certainly not standard is the elimination of ethics committees. Historically, much of the ethics work in healthcare has been done by ethics committees [13]. While non-acute facilities and some smaller hospitals may not have ethics committees, it is rare for larger hospitals to not have one. However, as the professionalization of clinical ethics continues to grow, several health systems with large ethics programs have closed their ethics committees, even at large quaternary care hospitals [14]. They did so because they find ethics committee members often lack the training to provide high-quality ethics consultation as compared to professional ethicists. Those who do have the training often do not have the time to staff an ethics consultation service given their other roles. Rather than committees, these hospitals rely on full-time professional ethicists for consultation, education, policy review, and other ethics activities. It remains to be seen if these closures are unique phenomena or the start of a growing trend. Either way, ethics consultation services will evolve significantly as the field grows.

Non-acute care settings present a significant opportunity for just this kind of growth. While acute care receives the most attention from clinical ethicists, the vast majority of patient encounters occur outside the walls of the hospital. As an example, in 2019, there were 36.5 million hospital admissions in the United States and just over 1 billion physician office visits [15, 16]. Yet there is comparatively little literature describing ethics consultation practices in non-acute areas or even discussing what it could or should look like [17–20]. The structure of these services challenges most traditional structures and models for ethics consultation. It is nearly impossible to have a truly representative ethics committee that covers a dozen sites in a network of clinics, and a professional ethicist would have a hard time being physically present at all those locations. Increasingly complex care is even being provided in patients' homes [21]. Ethics consultations services will have to adapt their structure, process, and model to be effective in these care settings. Clinical ethics cannot continue to ignore these patient encounters if it hopes to serve patients, their families, and clinicians.

Despite all this variation, some activities can be called a best practice even if they are not standard, including:

- The ethics consultant should document the consult in the patient's medical record.
- The ethics consultant should also document the consult in an internal tracking system kept by the ethics consultation service.
- The ability to request an ethics consult should be open to anyone involved in the case and should not be restricted to one group like physicians or nurses.
- Ethics consults should usually be performed by an individual ethics consultant or a small team of ethics consultants rather than an entire ethics committee.
- The attending physician should be made aware of the request for an ethics consult if they were not the one who requested it.

More descriptions of ethics programs and best practices are likely to emerge in the coming years, but it remains to be seen when or if consensus on process and practice will coalesce enough to truly call it a standard.

4.4 Impact of Clinical Ethics

The impact of clinical ethics can generally be divided into three categories: promoting the dignity of patients, their families, and clinicians; improving the quality of care; and providing operational benefits for the institution. First, clinical ethics aims to promote the dignity of every person involved including the patients, their family, and their clinicians [22, 23].[3] An ethics consult can help patients advocate for their needs, ensure that their voice is heard, identify and mitigate biases that might negatively affect their care, and support their rights when they conflict with other norms. For family members, an ethics consultant might walk with them as they make life and death decisions for their loved one, help them articulate their concerns or their understanding of the patient's wishes, and provide space for reflection on deeply held values. Regarding clinicians, an ethics consult can help support their rights like conscientious objection or to not be verbally abused, address their moral distress, and help face stressful situations. A clinical ethics consult provides someone to accompany them through some of the most trying professional encounters in their career. An ethics consult is one way the organization says to its caregivers, "We care about our moral health; we think value and integrity are important. Here is someone to help" [24]. An ethicist is "akin to a midwife, drawing forth the moral wisdom of" [25] the people involved and helping them articulate their perspectives. Helping

[3] I use dignity in the way Sulmasy describes "intrinsic dignity" or what Meilander calls the "coinherent dignity of our humanity and dignity of our person."

people express their moral values and be heard by others exhibits their intrinsic value as a person of equal moral worth; it promotes their dignity.

Second, clinical ethics can help improve the quality of clinical care. The Institute for Healthcare Improvement (IHI) defines quality as "the endeavor of continuously, reliably, and sustainably meeting customer [patient] needs" [26]. Clinical ethics fits this definition in several ways, in part by its focus on interpreting and applying the patient's wishes and values. For several decades, clinical ethics has focused significantly on the patient as the locus for medical decision-making. One goal of clinical ethics consults is to foster consensus on goals of care that are aligned, achievable, and in line with evidence-based medicine. Granted, consensus is not achievable in all cases, and an ethicist should not pursue consensus for the sake of consensus as it can undermine other more important considerations. Nevertheless, in my experience, many, if not most, clinical ethics consults result in better alignment of the care plan and the patient's value. Lastly, some requests from patients or surrogate decision-makers fall outside best practice, evidence-based medicine, or the standard of care. An ethics consult can help identify the scope of decision-making authority for patients and surrogates, the responsibility of clinicians to provide medically appropriate care, and their right (or obligation) to refuse requests that fall outside those boundaries. Thus, ethics consults can help strike a balance between these values to ensure that medical care meets quality standards and does not cause harm. At the same time, consults can foster flexibility from typical practice where possible when requested by a patient or surrogate, and help avoid care plans that are overly rigid.

Lastly, the question of operational and financial benefits of clinical ethics for the institution carries heightened importance given the ever-increasing focus on financial sustainability in healthcare. Many authors have written on this over the years, looking at everything from costs and length of stay (LOS) to the satisfaction of those who request ethics consults. There is no gold standard process, formula, or metric for analyzing the return on investment (ROI) for ethics consultation services [27].[4] More importantly, every hospital and health system focuses on different metrics to assess its own financial well-being. Some metrics are important in certain contexts and not important in others. This makes it difficult, if not impossible, for a single formula or metric to be meaningful for all ethics consultation services or leadership teams when assessing ROI. For example, almost all the literature on the ROI of clinical ethics has been published on hospitals in the United States. Given the unique payment structures in the United States, these metrics likely would not make sense (or may not even exist) in other countries. Even within the United States, a geography or health system with a primarily capitated model will focus on very different metrics than one with a primarily fee-for-service model. That being said, I have found three metrics to be particularly helpful when discussing ROI with healthcare leaders.

[4] For an excellent summary of the literature on published options for assessing the ROI of clinical ethics, see [27].

First is the variance between actual and expected LOS (hereafter LOS variance). Mark Repenshek first pioneered this approach by noting the difficulties with reviewing LOS broadly speaking [28]. Even a very high volume ethics consultation service will never have enough consults to impact the overall length of stay of a hospital or a unit; there are just too many patients and too few ethics consults to have that much of an impact. Instead, Repenshek proposed to stratify the variance between actual LOS and expected LOS by the time between admission and consultation. He identified expected LOS through a risk-adjusted model that matches patients based on diagnosis-related groups (DRGs). His data on 4 years of ethics consults show that earlier requests for an ethics consult have lower LOS variance. While his results are not statistically significant, they reflect the intuition of many ethicists: earlier consults involve less conflict and entrenchment of positions, which results in more consensus around goals of care and discharge plans. He only analyzed ICU patients to limit confounding variables like obstetrics patients or those with a generally short LOS.

Second, also described by Repenshek in the same article, is the difference between the observed and expected readmission rates. With a similar risk-adjusted methodology for expected readmission rates, based on DRGs, his data showed that ethics consults were associated with lower readmission rates when compared to patients with the same DRGs. He again looked only at patients in the ICU. For 2 of the 4 years, the difference in readmission rate was statistically significant. Ethics consults lowered the readmission rate by 8–10%.

Third is the difference in net margin for consults requested in the first 48 hours after admission and those requested more than 48 hours after admission. Multiple articles have analyzed the costs, charges, and revenue for ethics consults [29, 30].[5] Instead, Mary Homan turned to net margin because these other metrics do not accurately describe how much income the hospital actually received [31]. She defines net margin as "the difference between total costs and total payments" [31].[6] Over 4 years of ethics consults for Medicare patients, she found that patients who received an ethics consult within 48 hours of admission had a net margin $14,191.74 higher than those who received an ethics consult more than 48 hours after admission. Taken together, these three metrics have proven helpful in showing hospital leaders the financial benefit of ethics consultation services. Other metrics are also helpful, but in my experience, these metrics are the most meaningful for finance and operations leaders, at least in the United States. Unfortunately, no studies since these two have attempted to replicate these methodologies, so they are ideal for future research. To be clear, the benefit of ethics consultation services cannot be reduced to its financial impact; however, a good ethics consultation service cannot ignore this reasoning to advocate for the resources it needs to be effective.

[5] Two examples are Heilicser et al. [29] and Gilmer et al. [30].

[6] Homan [31, p. 89]. A similar metric, contribution margin, removes the indirect costs (overhead) to describe only the revenue that is impacted by the decisions of the care team and managers.

4.5 Conclusion

Clinical ethics is in a pivotal stage of development as it evolves into a field in its own right. Practice currently varies widely, and few things can truly be considered a standard. However, as best practices emerge in the coming years, some standards are likely to evolve. Given the morally pluralistic nature of society, standards will probably focus more around the structure of ethics programs and process of ethics activities (like consultation and data analysis) than the value-laden content of ethics advice. Such standards are vital for the growth of the field and for healthcare as a whole. Clinical ethics exists to promote and defend the human dignity of patients, their families, and clinicians. It stands as a fundamental bulwark in an increasingly technocratic, depersonalizing world. Standard practices will allow for better comparisons between ethics programs and better evaluation of a program's quality, which will in turn further humanize healthcare as a ministry.

References

1. Clinical Ethics Consultation Benchmarking Collaborative. n.d. https://www.cecbc.net. Accessed 21 Jun 2025.
2. Fox E, Danis M, Tarzian AJ, Duke CC. Ethics consultation in US hospitals: a national follow-up study. Am J Bioeth. 2022;22:5–18.
3. HCEC Certification Commission. HCEC Certification. n.d. https://asbh.org/certification/hcec-certification. Accessed 21 Jun 2025.
4. Council on Program Accreditation for Clinical Ethics Training (COPACET). n.d. https://fox-ethics.com/copacet-1. Accessed 21 Jun 2025.
5. Repenshek M. Hiring clinical ethicists: building on Gremmels' staffing model approach. Health Care Ethics USA. 2021;29(1):24–7.
6. Bruce CR, Lapointe J, Koch P, Lee K, Fedson S. Building a vibrant clinical ethics consultation service. Natl Catholic Bioethics Q. 2018;18:34.
7. Kenney MR. A system approach to proactive ethics integration. Natl Catholic Bioethics Q. 2018;18(1):93–112.
8. American Society of Anesthesiologists. Statement on ethical guidelines for the anesthesia care of patients with do-not-resuscitate orders. 2023. https://www.asahq.org/standards-and-practice-parameters/statement-on-ethical-guidelines-for-the-anesthesia-care-of-patients-with-do-not-resuscitate-orders.
9. Chidwick P, Bell J, Connolly E, Coughlin MD, Frolic A, Hardingham L, Zlotnik Shaul R. Exploring a model role description for ethicists. HEC Forum. 2010;22(1):31–40.
10. Gremmels B. Revising Repenshek's minimum standard set of data. Health Care Ethics USA. 2022;30(1):8–10.
11. Wasson K, Parsi K, McCarthy M, Siddall VJ, Kuczewski M. Developing an evaluation tool for assessing clinical ethics consultation skills in simulation based education: the ACES project. HEC Forum. 2016;28:103–13.
12. Sawyer KE, Dundas N, Snyder A, Diekema DS. Competencies and milestones for bioethics trainees: beyond ASBH's healthcare ethics consultant certification and core competencies. J Clin Ethics. 2021;32:127–48.

13. Post LF, Blustein J. Handbook for healthcare ethics committees. 3rd ed. Baltimore: Johns Hopkins University Press; 2021.
14. Mabel H, Crites JS, Cunningham TV, Potter J. Reimagining thriving ethics programs without ethics committees. Am J Bioeth. 2025;25:55–70.
15. American Hospital Association. Fast facts on U.S. hospitals, 2019. 2020. https://www.aha.org/statistics/2020-01-07-archived-fast-facts-us-hospitals-2019.
16. Centers for Disease Control and Prevention. National Ambulatory Medical Care Survey: 2019 National Summary Tables. 2019. https://www.cdc.gov/nchs/data/ahcd/namcs_summary/2019-namcs-web-tables-508.pdf.
17. Sanders A. Ethics across the continuum of care. In: Catholic health care ethics: a manual for practitioners. National Catholic Bioethics Center; 2020.
18. Barina R. Ethics outside of inpatient care: the need for alliances between clinical and organizational ethics. HEC Forum. 2014;26(4):309–23.
19. Burda ML. Exploring the next frontier: clinical ethics for the physician's office. Health Care Ethics USA. 2014;9(2):24-30.
20. Carrese JA, McDonald EL, Moon M, Taylor HA, Khaira K, Beach MC, Hughes MT. Everyday ethics in internal medicine resident clinic: an opportunity to teach. Med Educ. 2011;45:712–21.
21. AMA News Wire. Why is home health care growing? No one wants to be in a hospital. 2024. https://www.ama-assn.org/practice-management/digital-health/why-home-health-care-growing-no-one-wants-be-hospital.
22. Sulmasy D. Dignity and bioethics: history, theory, and selected applications. In: Lanigan B, editor. Human dignity and bioethics. Nova Science Publishers; 2008. p. 327.
23. Meilander G. Neither beast nor god: the dignity of the human person. New York: New Atlantis Books; 2009.
24. Payne K. Reflections on the role of ethicists in the Catholic health ministry. Health Care Ethics USA. 2010;18(2):26.
25. Kockler NJ, Dirksen KM. Integrating ethics services in a catholic health system in Oregon. Natl Catholic Bioethics Q. 2018;18:113–34.
26. Sampath B. What does it mean to rethink quality? 2021. https://www.ihi.org/insights/what-does-it-mean-rethink-quality.
27. Repenshek M. Examining quality and value in ethics consultation services. Natl Catholic Bioethics Q. 2018;18(1):59–68.
28. Repenshek M. Assessing ROI for clinical ethics consultation services. Health Care Ethics USA. 2017;12:12–20.
29. Heilicser BJ, Meltzer D, Siegler M. The effect of clinical medical ethics consultation on healthcare costs. J Clin Ethics. 2000;11(1):31–8.
30. Gilmer T, et al. The costs of nonbeneficial treatment in the intensive care setting. Health Aff. 2005;24(4):961–71.
31. Homan ME. Factors associated with the timing and patient outcomes of clinical ethics consultation in a Catholic health care system. Natl Catholic Bioethics Q. 2018;18:71–92.

Chapter 5
The Training of Clinical Bioethicists: Technical and Human Competencies Through Empathy

David Cerdio-Domínguez and Jonathan McFarland

5.1 Introduction

5.1.1 The Urgent Need for Clinical Bioethicists in Today's World

In a global context marked by unprecedented scientific advancements and increasingly complex ethical dilemmas, clinical bioethics has emerged as an essential field to guide healthcare practices toward a more humanistic approach [1]. Clinical bioethics, understood as the discipline that reflects upon, analyses, and proposes ethical solutions to issues arising in medical care and in the relationship between healthcare professionals and patients, has become a key tool in addressing the challenges of the twenty-first century [2, 3]. Since its inception in the 1970s, bioethics has evolved beyond mere theoretical reflection, requiring practices that integrate technical expertise with a profound understanding of the human condition [4].

While technological and scientific advancements have significantly improved clinical outcomes, they have also given rise to new ethical dilemmas: decisions regarding the beginning and end of life, the equitable distribution of resources, informed consent in vulnerable populations, and the risk of dehumanization in professional practice [5]. Within this landscape, the clinical bioethicist plays a crucial

D. Cerdio-Domínguez (✉)
Anahuac Centre for Strategic Development in Bioethics, Mexico City, Mexico

Anesthesiology Department, Regional Hospital Lic. Adolfo López Mateos ISSSTE, Mexico City, Mexico
e-mail: david.cerdio@anahuac.mx

J. McFarland
University of Pompeu Fabra, Barcelona, Spain

University of Autonomous Madrid, Madrid, Spain

D. Cerdio-Domínguez, M. E. de los Ríos-Uriarte (eds.), *Current Bioethics*, New Paradigms in Healthcare, https://doi.org/10.1007/978-3-032-25294-4_5

role, acting as a mediator and facilitator in ethical decision-making processes that directly impact the quality and dignity of healthcare [6].

The training of clinical bioethicists cannot be confined to the mere acquisition of theoretical knowledge or analytical skills [7]. While these competencies are essential, the true challenge for a bioethicist lies in their holistic human formation [8]. Those engaged in clinical bioethics must be capable of integrating ethical, legal, and medical knowledge with an empathetic sensitivity, allowing them to recognize in others not merely a patient or a professional, but a human being endowed with dignity [9].

The integration of these competencies highlights the need to train not only experts in theoretical bioethics but also, above all, individuals with a humanistic vision who perceive healthcare as a space of encounter between human beings [8]. As thinkers such as Martin Buber [10] and Fernando Savater [11] have pointed out, recognizing the other as an equal, with inherent dignity, forms the foundation of a truly ethical relationship (see Fig. 5.1). And, only through this perspective can clinical bioethics transcend the mere application of principles [12] and become a genuine instrument for strengthening humanism in the health sciences.

This chapter explores the importance of training clinical bioethicists from a dual perspective: technical excellence and the development of human capacities [13], with a particular emphasis on clinical empathy. We argue that for bioethical practice to be truly effective, it must be led by professionals who see human interaction as the core of their work, thereby achieving an authentic humanistic approach to healthcare. Through a review of fundamental philosophical and ethical concepts, this chapter seeks to demonstrate that a comprehensive training of clinical bioethicists is essential for addressing contemporary ethical challenges and transforming medical practice into a profoundly human experience.

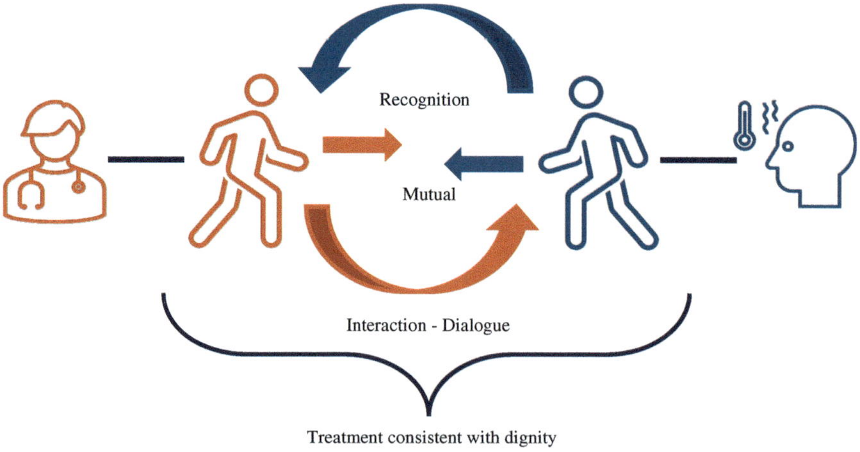

Fig. 5.1 Model of recognition of otherness and human dignity. (Own authorship)

5.2 Body

5.2.1 What Is Clinical Bioethics?

Clinical bioethics is a branch of bioethics dedicated to analyzing, addressing, and resolving ethical dilemmas that arise in clinical practice and focuses on decisions that directly impact patients' lives, health, and dignity [1, 14]. It emerged in response to growing concerns over the advancement of medical technology and its effect on human life, as well as historical abuses in clinical research that highlighted the need to regulate and protect the rights of individuals involved in healthcare [15]. Clinical bioethics has developed as an interdisciplinary field in this context, integrating elements from medicine, philosophy, law, and the social sciences [2, 16].

The early approach to clinical bioethics, influenced by figures such as Beauchamp and Childress [12], was characterized by formulating fundamental ethical principles, including autonomy, beneficence, non-maleficence, and justice. Although these principles were developed within a theoretical framework, they were intended as practical tools to guide decision-making in clinical contexts that demanded a pragmatic orientation [17]. However, other scholars, such as Edmund Pellegrino, emphasized the need to incorporate a more humanistic and relational dimension into clinical bioethics. He argued that the essence of this discipline should not be limited to applying general principles but must be rooted in promoting the patient's well-being and a profound respect for their vulnerability [18]. Likewise, authors such as Elio Sgreccia have highlighted the intrinsic value of the human person, advocating for bioethics that recognizes human dignity as the central foundation of all ethical action [19].

Much reflection has been devoted to bioethics' various worldviews and schools of thought [20]. However, regardless of the approach, methodology, or theoretical framework adopted (see Fig. 5.2), clinical bioethics must always maintain a single fundamental objective: protecting and promoting the human person, both the patient and the healthcare professional.

This commitment involves ensuring autonomy (or freedom) and safeguarding integrity and dignity, particularly in moments of extreme vulnerability, such as illness [21]. This unwavering focus on the human person makes clinical bioethics indispensable for fostering true humanism in the health sciences.

5.2.2 Current Needs and Relevance of Clinical Bioethics

Clinical bioethics has emerged as an essential response to modern medicine's complex phenomena and ethical challenges. Today, its relevance is heightened by various challenges beyond the technical aspects of healthcare and patient support, directly influencing the fundamental values of humanism in the medical sciences. These pressing needs not only justify the existence of clinical bioethics but also

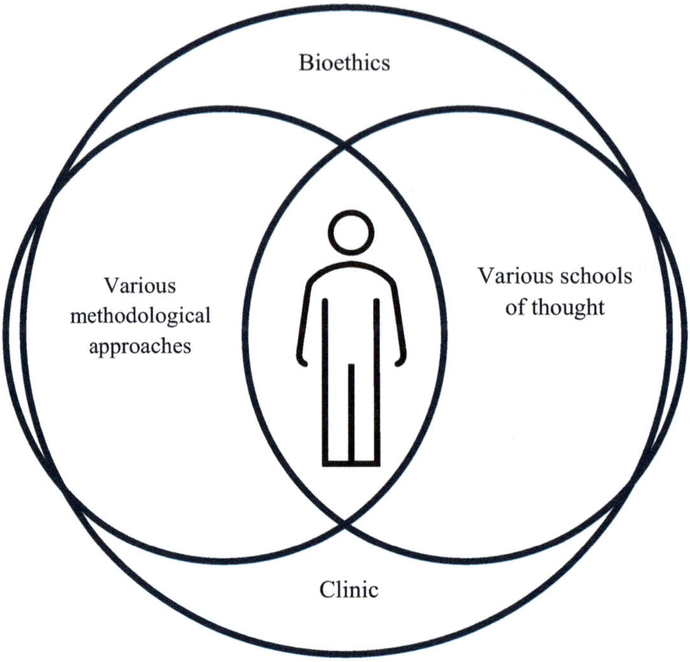

Fig. 5.2 The ideal objective of clinical bioethics. (Own authorship)

reinforce its role as a fundamental tool in ensuring and promoting truly dignified, person-centered care for both the patient and the healthcare professional.

One of the greatest challenges facing modern medicine is the tendency toward dehumanization, affecting healthcare professionals and patients alike [22]. This phenomenon largely stems from the predominance of a biomedical–technological approach, which prioritizes the disease over the person experiencing it. As William Osler aptly stated: "The good physician treats the disease; the great physician treats the patient who has the disease." In this context, the patient is often reduced to a mere collection of symptoms, disregarding their biological, psychological, social, and spiritual integrity [23, 24]. By focusing on human dignity and respect for the person, clinical bioethics seeks to counteract this reductionist perspective by promoting comprehensive care that considers the patient as a whole human being rather than merely an object of medical intervention. It establishes principles that facilitate ethical decision-making, particularly in moments of heightened vulnerability, where identifying the best course of action may be challenging. By grounding itself in the intrinsic value of the human person, clinical bioethics ensures that medical practice remains deeply committed to humanism and compassionate care.

The rapid advancement of medical technologies has profoundly transformed clinical practice, leading many healthcare professionals to focus primarily on technical proficiency in tools and procedures. While these innovations have significantly enhanced diagnostic and therapeutic capabilities, they have also fostered an overreliance on technology, often displacing ethical reflection in decision-making processes. Moreover, the fear of legal repercussions has contributed to the rise of "defensive medicine," where the protection of the physician takes precedence over patient well-being. In response to these challenges, clinical bioethics serves as a necessary counterbalance, emphasizing the primacy of ethical values in medical practice, beyond mere legal or technical compliance. By reinforcing human-centered care, bioethics ensures that medicine remains a profession guided by compassion, responsibility, and respect for human dignity.

Technological advancements have also given rise to complex ethical dilemmas surrounding the beginning and end of life, assisted reproduction, genetic editing, limiting life support, palliative care, and euthanasia. These issues raise profound questions about the meaning of life and death [25]. In this context, clinical bioethics becomes essential for addressing such challenges by providing a framework for ethical deliberation and informed decision-making. This approach ensures that medical decisions consider clinical indications and the values, beliefs, and wishes of patients and their families. By fostering transparent, compassionate, and ethically responsible decision-making, clinical bioethics helps to safeguard human dignity throughout these critical processes [26].

Beyond the previously mentioned challenges, clinical bioethics also confronts ethical dilemmas arising from the globalization of healthcare, such as unequal access to advanced treatments, clinical trials in developing countries, and healthcare delivery in humanitarian crises [27]. The COVID-19 pandemic exposed significant ethical tensions regarding the allocation of scarce resources, vaccination policies, and the prioritization of vulnerable populations [28]. These challenges highlight the need for a flexible clinical bioethics framework that responds to the demands of a rapidly changing world.

The ethical complexities facing modern medicine reinforce the importance of clinical bioethics as an essential field. It ensures that medical practice extends beyond technical problem-solving to actively promote humanism. From safeguarding patient dignity to facilitating ethical reflection in complex decision-making, clinical bioethics plays a crucial role in shaping a more just and compassionate healthcare system.

5.2.3 The Need for a Comprehensive Training in Clinical Bioethics

In today's world, where technological and scientific advancements have profoundly transformed clinical practice, the training of clinical bioethicists has become essential—not only to meet the technical demands of the field but also to address the ethical and human challenges that arise in daily medical care [7, 29]. It is therefore crucial to reflect on the need for a holistic education that enables bioethicists to develop as well-rounded individuals, capable of recognizing their own dignity and, through this self-awareness, identifying their peers in others [30].

Comprehensive training is based on the principle that human beings are not merely biological or technical entities but persons with biological, psychological, social, and spiritual dimensions [31]. Understanding this multidimensional nature is essential for a clinical bioethicist, as their role is not limited to theoretical analysis of ethical dilemmas but involves engaging in decision-making processes that directly impact people's lives.

This approach demands that the professional move beyond reductionist views that perceive the patient merely as a case or a problem, instead adopting a holistic perspective that places the inherent dignity of every human being at the center of ethical deliberation [32].

Dignity, as a universal criterion [33], is the fundamental axis guiding the doctor–patient relationship and ethical decision-making. This perspective, rooted in global ethical and anthropological principles, enables overcoming cultural and ideological barriers that may arise in diverse contexts [34]. A comprehensively trained bioethicist not only recognizes the dignity of others but also actively protects and promotes it, ensuring that every decision reflects a profound respect for the patient's vulnerability and humanity.

The concept of comprehensive training extends beyond the development of technical and theoretical competencies to include cultivating human skills such as empathy, active listening, and the ability to engage in meaningful dialogue [35]. These skills enable the bioethicist to establish genuine relationships based on trust and mutual respect, essential elements in creating a caring environment that facilitates informed and responsible ethical decision-making. In this sense, comprehensive training aims not only to produce experts in clinical bioethics but also to form individuals who embody and practice the values they seek to promote [6].

Throughout these pages, we have highlighted how clinical bioethics is at the core of human relationships in healthcare. A key question arises from this perspective: What does it mean to comprehensively train a clinical bioethicist? Answering this question requires considering the multiple dimensions of the human being and how they interact to shape a genuinely humanistic approach to bioethical practice. This inquiry encourages reflection on the essential elements that should be integrated into educational programs for these professionals, incorporating not only theoretical knowledge and technical skills but also values and attitudes that promote respect for human dignity in all its forms [13, 36].

Comprehensive training is not an abstract ideal, but a real necessity in a world where medical practice and ethical decision-making constantly evolve. Clinical bioethicists trained with this vision will be better equipped to navigate the ethical dilemmas of the twenty-first century, not only from a technical standpoint but also with a profound understanding of what it means to be human. This holistic preparation benefits patients and healthcare professionals, and it contributes to developing a more ethical, equitable, and humanistic healthcare system.

In the following sections, we will explore the specific components of this comprehensive training, including the technical and human competencies essential for clinical bioethicists to carry out their work with excellence and humanity.

5.2.4 Comprehensive Competency-Based Training for a Clinical Bioethicist

The role of a clinical bioethicist requires comprehensive preparation that encompasses not only technical knowledge but also practical skills and deeply human attitudes. To understand this training approach, first, it is essential to define competency as a set of technical, practical, and attitudinal skills (see Fig. 5.3) that enable an individual to perform effectively in a specific context [13, 37]. This holistic approach to training aims to ensure that bioethicists can address ethical dilemmas with academic rigor and human sensitivity. Since clinical bioethics does not merely operate within a regulatory framework but also seeks to integrate decision-making with a humanistic and ethical perspective, it is crucial that professionals develop specific skills. These competencies can be classified into three key dimensions: theoretical, practical, and attitudinal.

1. *Theoretical knowledge*
   Theoretical knowledge forms the foundation of a clinical bioethicist's work, enabling them to understand the ethical, legal, and cultural principles that guide their practice.
   (a) *Ethical fundaments*: Understanding ethical theories and bioethical principles (autonomy/freedom-responsibility, beneficence/defense of physical life, non-maleficence/therapeutic proportionality, and justice/solidarity) is essential [2, 3]. These principles are not merely conceptual tools but practical guidelines for navigating complex ethical and bioethical dilemmas [4].

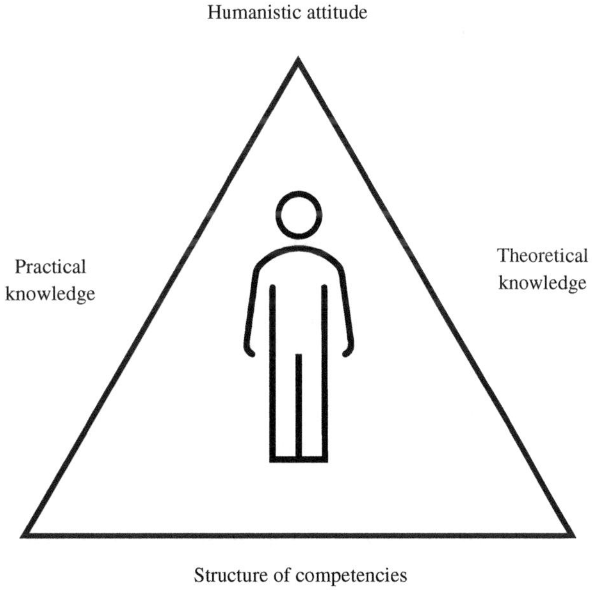

Fig. 5.3 Competency-based training. (Ref. [37]. Diseño: Cerdio-Domínguez, David; McFarland, Jonathan. Own authorship)

(b) *Legal and regulatory framework*: A clinical bioethicist must have a solid understanding of the laws and regulations that frame clinical decision-making, including patient rights, informed consent, and legal provisions regarding the beginning and end of life [38]. This knowledge ensures that ethical decisions align with the region's current legal frameworks, thereby promoting responsible and safe medical practice.
(c) Understanding cultural, religious, and social differences is crucial in clinical bioethics [36, 39]. Every patient brings a unique background that shapes their values and decision-making processes [20]. A clinical bioethicist must be able to analyze and respect these differences while promoting an approach that balances respect for diversity with the application of universal ethical principles [33].

2. *Practical skills*
Practical skills must complement theoretical knowledge, enabling bioethicists to apply their training in real and complex contexts, ensuring that bioethics is truly recognized as an applied discipline [40].

(a) *Evaluation and analysis*: A clinical bioethicist must be able to identify, analyze, and resolve ethical dilemmas in clinical settings. This requires using methodologies such as ethical deliberation, case analysis, and applying bioethical principles to specific scenarios [2].
(b) *Effective communication*: Engaging in dialogue with patients, families, and healthcare teams is essential. The bioethicist must be able to explain complex concepts in a clear and accessible manner, fostering a trust-based environment that facilitates shared decision-making [35].
(c) *Mediation and facilitation*: In many cases, the bioethicist mediates ethical conflicts between the parties involved. This skill requires impartiality, active listening, and the ability to guide participants toward consensus-based solutions that respect the values and needs of all stakeholders [41].

3. *Attitudes and values*
In addition to the theoretical and practical aspects of competencies, it is essential to reflect on the attitudinal component within the ethical and humanistic framework of a clinical bioethicist's training, ensuring that their practice aligns with the promotion and protection of human dignity [42].

(a) *Integrity and professionalism*: The bioethicist must uphold professional ethics, maintain confidentiality, and take responsibility for their decisions. Integrity is essential for building trust with patients and healthcare teams [43].
(b) *Empathy and respect*: Sensitivity to the perspectives and values of others is essential. The bioethicist must acknowledge the patient's vulnerability and approach them with compassion, fostering an environment where they feel heard and valued [44].
(c) *Critical reflection*: The bioethicist must continuously assess their practice, striving to enhance their ethical understanding and responsiveness to emerg-

ing challenges. This critical reflection enables ongoing development in an ever-evolving field [45].

4. *Clinical bioethics case—Competencies*
An elderly patient with advanced respiratory failure arrives at the emergency department. The family requests intubation to prolong the patient's life, while the medical team believes that intubation would only extend the patient's suffering. The patient, who is conscious, expresses their wish to refuse invasive interventions, a decision formally documented as an advance directive.

In this case, the clinical bioethicist must apply:

1. Theoretical knowledge:
   - They must apply the appropriate methodologies, starting with a clinical analysis based on the principles of (1) respect for physical life (beneficence), (2) globality and/or therapeutic proportionality (non-maleficence), (3) freedom-responsibility (autonomy), and (4) solidarity-subsidiarity (justice) to address the presented dilemma from an axiological perspective.
   - Additionally, they must take into account the legal regulations concerning the patient's advance directive.
2. Practical skills:
   - They will seek to facilitate dialogue between the family and the medical team, explaining the ethical and clinical implications of the decisions.
   - They will mediate to ensure that the patient's wishes are respected, while also addressing the family's emotional distress and promoting respect for the conscientious convictions of the treating medical team.
3. Attitudes and values:
   - They must approach the situation with clinical empathy toward the patient and their family, validating their feelings and concerns.
   - They will aim to maintain a thoughtful and professional approach, ensuring that the final decision is ethically sound and centered on the patient's dignity.

This example illustrates how theoretical, practical, and attitudinal competencies are integrated into the clinical bioethicist's practice, ensuring ethical, humane, and effective care.

5.2.5 To Be a Good Doctor, One Must First Be a Good Person

Beyond its technical and scientific complexity, the practice of medicine is deeply rooted in human interaction and the professional's ability to recognize and respect the dignity of each patient [46]. While universities and training programs often focus on developing technical and practical competencies, it is crucial to understand that excellence in healthcare does not solely depend on these skills but also a strong humanistic and ethical foundation. Simply put; to be a good doctor, one must first be a good person.

5.2.5.1 The Person at the Center of Medical Training

A doctor's training cannot be limited to the acquisition of clinical knowledge or the mastery of advanced procedures. While these skills are essential, they are insufficient to address the ethical and human challenges in everyday medical practice [47]. Healthcare, at its core, is an encounter between people: the doctor, with their knowledge and expertise, and the patient, with their vulnerability, dignity, and unique context. In this encounter, how the person is treated becomes the fundamental axis of every interaction [48].

From this perspective, a doctor should not only be an expert in diagnosis and treatment but also a facilitator of care who recognizes the inherent dignity of each patient [49]. This means the professional must be trained in ethical principles and humanistic values that guide their actions, thus responding to the call to act with empathy and compassion.

5.2.5.2 Dignity in the Clinical Encounter

Recognizing human dignity is the foundation upon which an ethical and effective doctor–patient relationship is built. This recognition is not an abstract concept, but a practical guide that should be reflected in every clinical interaction. A doctor who views their patients merely as "cases" or "problems" risks dehumanizing care, reducing medical attention to a purely technical exercise. In contrast, a professional who understands patient's dignity as a central value will be better equipped to establish a relationship of trust, which is essential for comprehensive care [8, 46].

Dignified treatment is also reflected in how the doctor approaches vulnerability. Every patient enters a consultation with fears, uncertainties, and unique needs. It is the doctor's responsibility to create an environment where the patient feels valued and heard. Moreover, doctors also experience vulnerability, whether due to emotional strain, ethical dilemmas, or the inherent limitations of medical practice. Reflecting on this dual vulnerability is essential to fostering a professional practice based on mutual understanding and respect.

5.2.5.3 The Role of Universities and Educators

Educational institutions play a fundamental role in training doctors who are technically and scientifically competent but also ethical and humanistic. Achieving this requires a shift in traditional teaching approaches, where the emphasis is placed not only on technical content but also, above all, on the comprehensive development of the individual. To achieve this goal, the following alternatives are proposed:

1. Curriculum centered on medical humanism
   - Incorporate subjects that promote ethical reflection, empathy, and respect for human dignity.
   - Encourage student participation in clinical activities that expose them to complex realities, fostering a deeper understanding of human vulnerability.
2. Role models
   - Professors and educators should serve as living examples of the values they seek to instill. Their conduct in the classroom and clinical environment directly influences how future doctors perceive and practice medicine.
3. Encouraging personal reflection
   - Create spaces where students can reflect on their own motivations, values, and attitudes toward healthcare.
4. The impact on healthcare
   - A doctor who understands the importance of being a good person enhances their professional practice and contributes to the transformation of the healthcare system. A person-centered approach can reduce levels of dehumanization and promote more equitable and just care.
   - Moreover, this approach benefits the patient and the professional, creating a care environment where both feel valued and respected.

The path toward ethical and humanistic medicine begins with the training of doctors who, before being technical professionals, are good human beings. This requires recognizing that dignified treatment, empathy, and respect for vulnerability are as essential as clinical knowledge. Universities and educators are responsible for creating spaces that foster this comprehensive development, ensuring that future doctors can face the challenges of modern medicine with humanity and professionalism. The true success of a doctor is not measured solely by accurate diagnoses but by their ability to touch lives with dignity and compassion.

5.2.6 *Humanism and the Humanities in Health Sciences*

The concept of humanism in health sciences goes beyond technical and scientific competencies, requiring healthcare professionals to recognize and respect the inherent dignity of every person [50]. This section explores how the humanities can foster a profound sense of humanity in doctors, promoting an "ethical mirror" in which, by recognizing themselves as individuals, they can fully acknowledge others as equals [31].

5.2.6.1 The Humanities and the Sense of Humanity

The humanities—including philosophy, literature, and art, among other disciplines—play a fundamental role in developing ethical sensitivity in healthcare professionals, as they help to:

1. Foster empathy: Enable doctors to understand human experiences of vulnerability, suffering, and dignity.
2. Encourage self-reflection: Provide a space for doctors to question their role, values, and relationship with the patient.
3. Recognize the other: Invite healthcare professionals to see the patient not as a clinical object but as a whole person with emotions, fears, and hopes.

5.2.6.2 The "Mirror Law" in the Doctor–Patient Relationship

A doctor can only recognize their patient's dignity if they first acknowledge their dignity as a human being. This creates a reflective effect, where the shared vulnerability between doctor and patient becomes a bridge to empathy. The doctor can offer genuine care by seeing the patient as an equal. In line with this perspective, Kipling's poem "If" is proposed as a starting point for further reflection on the personal and professional development of doctors.

If
By Rudyard Kipling

If you can keep your head when all about you
Are losing theirs and blaming it on you;
If you can trust yourself when all men doubt you,
But make allowance for their doubting too:
If you can wait and not be tired by waiting,
Or being lied about, don't deal in lies,
Or being hated, don't give way to hating,
And yet don't look too good, nor talk too wise;

If you can dream-and not make dreams your master;
If you can think-and not make thoughts your aim;
If you can meet with Triumph and Disaster
And treat those two impostors just the same:
If you can bear to hear the truth you've spoken
Twisted by knaves to make a trap for fools,
Or watch the things you gave your life to, broken,
And stoop and build 'em up with worn-out tools;

If you can make one heap of all your winnings
And risk it on one turn of pitch-and-toss,
And lose, and start again at your beginnings
And never breathe a word about your loss:
If you can force your heart and nerve and sinew
To serve your turn long after they are gone,

And so, hold on when there is nothing in you
Except the Will which says to them: 'Hold on!'

If you can talk with crowds and keep your virtue,
Or walk with Kings-nor lose the common touch,
If neither foes nor loving friends can hurt you,
If all men count with you, but none too much:
If you can fill the unforgiving minute
With sixty seconds' worth of distance run,
Yours is the Earth and everything that's in it,
And-which is more-you'll be a Man, my son!

When integrating it into the context of medical humanism, the following key points can be highlighted:

1. Self-control and resilience: The ability to "keep your head when all about you are losing theirs" reflects the importance of emotional balance in complex clinical situations. How would we like our doctor to behave? How are we conducting ourselves?
2. Empathy without prejudice: "If you can meet with Triumph and Disaster and treat those two impostors just the same" highlights the need to avoid judgment and not overvalue success or failure in medical care, focusing on the patient rather than reducing them to clinical outcomes.
3. Commitment to continuous effort: "If you can fill the unforgiving minute with sixty seconds' worth of distance run" serves as a reminder that every moment matters in the doctor–patient relationship. Physicians should fully commit to each interaction, treating patients under the premise: "treat others as you would like to be treated."

5.2.7 Integration of the Humanities in Medical Education

To ensure that doctors develop a profound sense of humanity, universities must integrate the humanities as an essential component of their educational programs:

1. Ethics and philosophy courses: Encourage critical reflection on human dignity and values.
2. Literature and arts: Help doctors understand human experiences from a broader perspective.
3. Comprehensive training programs: Promote both technical expertise and humanistic development.

Humanism in health sciences is not an optional trait but an essential component for ensuring ethical and dignified care. Through the humanities, doctors can develop a comprehensive vision that fosters mutual respect, empathy, and the full recognition of the humanity of others. As Kipling suggests, true greatness lies in maintaining virtue, resilience, and humanity in every clinical interaction.

5.3 Conclusion

The comprehensive development of clinical bioethicists is fundamental in addressing the ethical challenges of contemporary medicine. This chapter highlights that bioethical practice is not merely about mastering theories and ethical principles but requires a deep commitment to human dignity and humanism. In a context where dehumanization and the dominance of technology in healthcare are growing concerns, bioethicists are called to facilitate dialogue, fostering a genuine connection between ethical values and the practical needs of both patients and professionals.

To achieve this goal, clinical bioethics training must focus on three interrelated dimensions:

1. Theoretical Knowledge: Understanding bioethical principles, legal regulations, and cultural diversity provides a solid foundation for ethical deliberation.
2. Practical Skills: The ability to analyze dilemmas and facilitate dialogue in ethical conflicts is essential to ensure that decisions uphold human dignity.
3. Attitudes and Values: Empathy, integrity, and critical reflection are indispensable qualities for promoting an ethical practice that acknowledges both the vulnerability of others and one's own.

This chapter highlights that being a competent clinical bioethicist requires more than just technical expertise—it demands a profound recognition of the intrinsic value of the human person. This humanistic approach must be integrated into educational curricula, balancing technical proficiency and ethical reflection. Only through this holistic training can the doctor–patient relationship be strengthened, ethically responsible decisions be ensured, and sensitive responses be developed to address twenty-first-century challenges, such as the equitable distribution of resources, the globalization of healthcare, and end-of-life dilemmas.

In summary, clinical bioethics, as an interdisciplinary and humanistic discipline, goes beyond solving ethical dilemmas. It seeks to transform medical care into a deeply human experience centered on the dignity of all those involved. This commitment to humanism and empathy is the essence of genuine bioethical practice.

References

1. Baker R. Bioethics and history. J Med Philos. 2002;27(4):447–74.
2. Kuthy Porter J, Villalobos Pérez JJ, Martínez González ÓJ, Tarasco Michel M. Introducción a la Bioética. 4th ed. México: Universidad Anáhuac México; 2015.
3. Sgreccia E. Manual de Bioética. México: Editorial Diana; 1996.
4. Emmerich N. What is bioethics? Med Health Care Philos. 2015;18(3):437–41.
5. Félix J, Cerdio D, Del Campo E, Gutiérrez R, Castro L, Cedillo A. Relación médico-paciente e inteligencia emocional, un reto en la educación médica. Medicina y Ética. 2021;32:635–50.
6. Degrazia D. Value theory, beneficence, and medical decision-making. Am J Bioeth. 2020;20(3):71–3.

7. American Society for Bioethics and Humanities. Core competencies for healthcare ethics consultation. 2nd ed. Chicago: ASBH; 2011.
8. De los Ríos-Uriarte ME, Cerdio-Domínguez D, Alfredo-González J, Ricaud-Velez IA. Fundamentos Antropológicos y Éticos de la Relación Médico-Paciente y su Dinámica durante la Pandemia por COVID19. Cuba: Bioética: Instituto de Bioética Juan Pablo II; 2022.
9. Ortiz-Millán G. La dignidad, entre el escepticismo y el entusiasmo. Instituto de Investigaciones Filosóficas: Universidad Nacional Autónoma de México, Colegio de Bioética; 2014. p. 1–16.
10. Buber M. ¿Qué es el Hombre? FCE; 1967.
11. Savater F. Ética para Amador. 1st ed. Barcelona: Ariel; 2022.
12. Beauchamp T, Childress J. Principles of biomedical ethics: marking its fortieth anniversary. Am J Bioeth. 2019;19(11):9–12.
13. Cerdio Domínguez D, de Los Ríos Uriarte ME, Molina de la Rosa AA, Félix Castro JM, Fueyo Minutti A, Del Campo Turcios EC, et al. Medical theatre as an innovative pedagogical strategy in medical simulation. Med Sci Educ. 2022;32:719–22.
14. Beauchamp TL. Methods and principles in biomedical ethics. J Med Ethics. 2003;29(5):269–74.
15. Ten Have HA. Potter's notion of bioethics. Kennedy Inst Ethics J. 2012;22(1):59–82.
16. CONBIOÉTICA. ¿Qué es la bioética? Comisión Nacional Mexicana de Bioética; 2015. p. 263–77.
17. Veatch RM. Reconciling lists of principles in bioethics. J Med Philos. 2020;45:540–59.
18. Pastor LM. From virtue bioethics to bioethics personalistic: is integration possible? Cuad Bioet. 2013;24(80):49–56.
19. Sgreccia E. Personalist bioethics. Philadelphia: National Catholic Bioethics Center; 2012.
20. Cerdio-Domínguez D, De los Ríos-Uriarte ME, García-Llaca E. Cosmovisiones en Bioética, interpretación del dolor y el sufrimiento. Apuntes de Bioética [Volumen y páginas no especificados], Universidad Católica Santo Toribio de Mogrovejo; 2023.
21. Ferreira JBB, Santos LLD, Ribeiro LC, Rodrigues Fracon BR, Wong S. Vulnerability and primary health care: an integrative literature review. J Prim Care Community Health. 2021;12:215013272110497.
22. Barstow C, Shahan B, Roberts M. Evaluating medical decision-making capacity in practice. Am Fam Physician. 2018;98:40–6.
23. Francis. Laudato SI': Carta encíclica del Sumo Pontífice Francisco. Iglesia Católica; 2015.
24. Burgos JM. Antropología Breve. España: Ediciones Palabra; 2010.
25. Cerdio Domínguez D. El manejo del dolor crónico -no oncológico-: un reto bioético emergente. Medicina y Ética. 2022;33:505–25.
26. Gómez-Álvarez JE. Elio Sgreccia: una bioética ontológicamente fundada, Edición Especial. CISAV; 2023.
27. Ten Have H. The challenges of global bioethics. Glob Bioeth. 2022;33(1):41–4.
28. Llaca-García E, Templos LA. La pandemia de COVID-19 en México: el papel fundamental de los cuidados paliativos y bioética. In: Templos-Esteban LA, editor. Medicina y Ética, vol. 32. México: [Editorial no especificada]; 2021. p. 179–95.
29. Bishop CE, Maradiaga G, Freeman KR, Peters TR, Jackson JM. Simulation: an innovative approach to engaging preclinical medical students with bioethics. Med Sci Educ. 2021;31(2):325–9.
30. Cerdio Dominguez D, Millán Zurita P, Cedillo Urbina AC, Félix Castro JM, Del Campo Turcios EC. Social intelligence, an elementary competence in the development of the doctor-patient relationship. Proc Sci Res Univ Anáhuac Multidisci J Healthcare. 2021;1(1):52–61.
31. Cerdio-Domínguez D. Spirituality as the basis and Foundation of the Medical Profession. 1st ed. Springer Nature; 2022.
32. Kalkach Aparicio M, Carrillo Ruíz JD. Neuroética. 1st ed. México: Manual Moderno; 2020.
33. UNESCO. Declaración Universal sobre Bioética y Derechos Humanos. UNESCO; 2005.
34. Macpherson C. Global bioethics: it's past and future. Glob Bioeth. 2022;33(1):45–9.
35. Forguione-Pérez VP. Comunicación entre médico y paciente: más allá de una consulta, un proceso educativo. Médicas UIS. 2015:7–13.

36. Ten H, Gordijn B. Handbook of global bioethics. Springer Science + Business; 2014.
37. ASBH, editor. Core competencies for healthcare ethics consultants. 3rd ed. ASBH; 2022.
38. Hervada J. Introducción Crítica al Derecho Natural. 11th ed. España: Ediciones Universidad de Navarra; 2011.
39. OMS. The Global Health Observatory. OMS; 2020.
40. Souza LA, Pessoa AP, Barbosa MA, Pereira LV. The bioethical principlism model applied in pain management. Rev Gaucha Enferm. 2013;34(1):187–95.
41. Guerra López R. Bioethics and rationality. Personalism at the service of expanding the horizon of reason at the foundation of bioethics. Cuad Bioet. 2013;24(80):39–48.
42. Berger AS, Niedra E, Brooks SG, Ahmed WS, Ginsburg S. Teaching professionalism in postgraduate medical education: a systematic review. Acad Med. 2020;95(6):938–46.
43. Lindberg C, Fagerström C, Sivberg B, Willman A. Concept analysis: patient autonomy in a caring context. J Adv Nurs. 2014;70(10):2208–21.
44. Dow AW, Leong D, Anderson A, Wenzel RP, Team VCUT-M. Using theater to teach clinical empathy: a pilot study. J Gen Intern Med. 2007;22(8):1114–8.
45. Richards JB, Hayes MM, Schwartzstein RM. Teaching clinical reasoning and critical thinking: from cognitive theory to practical application. Chest. 2020;158(4):1617–28.
46. Advices from Aesculap to his medical son. In.
47. Macpherson I, Roqué MV, Martín-Sánchez JC, Segarra I. Analysis in the ethical decision-making of dental, nurse and physiotherapist students, through case-based learning. Eur J Dent Educ. 2022;26(2):277–87.
48. Gracia D. Entrevista a Henk ten Have. EIDON. 2014:53–74.
49. Fernández-Varela Mejía H, Sotelo Monroy GE. El deber del cuidado médico en México. Rev Facult Med. 2019;62(3):40–9.
50. Hwang I. Medical humanism in the posthuman era: a critical examination of its past, present, and future. Uisahak. 2023;32(1):115–45.

Chapter 6
Clinical Bioethics in Latin America: Current State

María Bernardita Portales

6.1 Introduction

In the daily clinical care of patients and users, ethical problems often arise, presenting situations that cause doubt and uncertainty. These situations involve conflicting values and require clinical ethics reflection to find the best decisions for each individual. Such problems may occur at all levels of healthcare complexity, not just in large hospitals or tertiary care settings, appearing in any clinical encounter. This is where clinical ethics (also known as *clinical bioethics* in Spanish-speaking context), understood as the branch of bioethics applied to clinical practice, becomes fundamental. It guides decisions and actions when doubts or disagreements are generated among those involved. Thus, clinical ethics is linked to professional exercise, focusing on the practical, specific decisions that raise questions about values in the professional–patient relationship, among treating teams, and within healthcare institutions [1].

Clinical ethics is expressed in everyday healthcare practice and does not aim for universality [2]. Its scope of action lies in the relationship between the specific person seeking care (and their family) and the professionals attending to them. "Doing clinical bioethics implies a direct approach to the patient, a sustained dialogue with them, and an analysis of their circumstances" [3]. Consequently, effective communication and interpersonal relationships among all agents involved in decisions—doctors, other health professionals, patients, and families—are of fundamental importance [1]. In scenarios of uncertainty and questioning, this disciplinary area seeks, through reflection and deliberation, reasoned and prudent courses of action focused on what is best for each particular patient, considering their reality and

M. B. Portales (✉)
Center for Bioethics, Faculty of Medicine, Clínica Alemana Universidad del Desarrollo, Santiago de Chile, Chile
e-mail: bportales@udd.cl

D. Cerdio-Domínguez, M. E. de los Ríos-Uriarte (eds.), *Current Bioethics*, New Paradigms in Healthcare, https://doi.org/10.1007/978-3-032-25294-4_6

context. It is therefore essential to understand the values, criteria, and preferences of the person seeking care, as well as their social, cultural, and family context.

In Latin America, bioethics, across its various areas of development, has been characterized by an interdisciplinary and local perspective. This gives it shared characteristics, but also certain differences due to the heterogeneity of local realities. The region comprises countries with unequal levels of development, each with its own qualities and particularities [4], history, culture, and population characteristics. These factors pose unique challenges and reflections that must be considered. For authors like Luca Varela and Leo Pessini, compared to bioethics originating in the United States, bioethics in Latin America has a more social focus or responsibility. This has led Latin American bioethics to concentrate on community problems taking into account social and public components where values such as justice, equity, and solidarity gain relevance. Concrete examples include issues related to access to healthcare and the defense of the rights of vulnerable people [5].

Clinical ethics in Latin America develops within its own regional context, marked by social diversity, varied traditions, multiculturalism, religiosity, and spirituality, all within a context of moral pluralism. Clinical relationships between professionals and patients also have their own characteristics, influenced by the aforementioned differences, as well as by the specific reality (time, place, and context) of the encounter. While healthcare contexts may vary across countries and localities, and bioethical problems will arise from practices in concrete and specific situations and realities [3], there are actions that demonstrate the development and institutionalization of clinical ethics in Latin American countries. Although the development of bioethics in the Latin American context may still be limited in certain practical areas [6], there are indicators reflecting significant positive advances. These include the formation of healthcare ethics committees, the implementation of bioethics training programs, and the creation of national bioethics commissions [7].

This chapter aims to present the current state of clinical bioethics in Latin America by identifying recognized clinical ethics problems in healthcare and describing the establishment of clinical ethics committees, setting out examples of regulations that govern them and discuss challenges in their operation and recognition. For the proper development of clinical ethics, education in this discipline is undoubtedly necessary; this topic is addressed in other chapters of this book and will only be commented on here in relation to the educational function of these committees.

6.2 Clinical Ethics Decisions in Healthcare Practice

Healthcare professionals make decisions daily. In the care of ill people, patients, and users [8], whether in primary care or hospitalization settings, these decisions can be complex, generating numerous questions and challenges that require deliberation and ethical reflection, considering specific circumstances and identifying involved values.

In Latin America, as in many other regions, patients have increasingly been recognized as a moral agent, leading to respect for their autonomy and participation in health decision-making. Medicine has transitioned from paternalistic clinical relationship models to those promoting patient participation, resulting in shared decision-making. While differences in this transition may exist in Latin America, as a study by Lazcano et al. shows that paternalism is still present in some countries [9], legislation has advanced to recognize patients as moral agents in health decisions. This reflects the growing importance of this recognition, as norms and laws reflect the values prioritized in different cultures [10].

Respect for patient autonomy constitutes the main ethical foundation of informed consent, a crucial process for patients to exercise autonomy and increasingly relevant in current clinical relationships. For autonomous decisions, patients need accurate and sufficient information, delivered in understandable language, with content depth appropriate to the decision being discussed [11]. Given Latin America's sociocultural diversity and unique history, fostering patient participation may also require cultural adaptation to align the clinical relationship with each location's and healthcare center's idiosyncrasies. Moreover, this relationship should promote values such as respect for ontological dignity, nondiscrimination and non-stigmatization, tolerance and solidarity, as well as principles of privacy, confidentiality, equality, equity, and human rights [12]. Interestingly, a study by García et al. identified "ethical aspects of the patient–provider relationship" as one of four emerging themes in Latin American bioethics publications, alongside "questions involving the beginning and end of life" (e.g., embryo status, abortion, palliative care, and end-of-life care), "ethical education and training of health professionals," and "ethics in research with human subjects" [13]. Other publications note ethical problems related to the clinical relationship, respect for autonomy and confidentiality, rights of older adults, children, and adolescents, rights of migrants [10, 14], and challenges posed by telemedicine in Latin America [15].

In addition to changes in the clinical relationship, medicine has seen significant scientific advances, leading to progress in diagnoses and treatments. In contemporary medicine, this development is accompanied by overspecialization and technification, which carries the risk of patient fragmentation and the dehumanization of healthcare. There's a danger of forgetting the patient's unique sociocultural context, their health and illness experiences, and their diverse expectations and fears about healthcare. While highly technified and specialized medicine offers undeniable benefits, it is crucial to remember the need for a patient-centered medicine that focuses on the individual and their needs, not just the disease, diagnosis, or affected organ. Montserrat Esquerda emphasizes that in this era of great progress, "professionals are required who not only master technical knowledge, but are also capable of adapting to new social demands, respecting cultural diversity, and maintaining the balance between science, ethics, and humanity" [16].

We must not forget that decisions in healthcare practice are made in contexts of uncertainty, increasing complexity and raising questions that demand ethical reflection and justification of different opinions and courses of action. New questions and challenges for reflection emerge in contemporary medicine. We must consider the limits and purposes of medicine, whether everything technically possible is ethically correct, and deliberate on the best decisions in uncertain situations, "ensuring that the patient's well-being is the priority, thus promoting a more human and respectful healthcare environment" [1]. The voices of all involved, especially patients and their families, must be heard, remembering that in a morally pluralistic society, each person has their own scale of values. A balance must be struck between respecting patient's autonomy—a relational autonomy influenced by their social connections—and the responsibility of healthcare professionals to provide the best possible care. To achieve collaborative clinical relationship models that facilitate shared decision-making in various clinical encounters, respecting and centering on the patient as a person, healthcare professionals also need to develop and acquire attitudes, behaviors, and virtues associated with virtue ethics and care ethics [1].

As Patricio Silva notes, "clinical practice is subject to increasingly demanding challenges" [17]. Beyond the scientific and technical complexity of many decisions, it is always necessary to consider the "social and economic factors in the particular reality of each patient and each family" [18]. This also means keeping in mind, in Latin American clinical care, "cultural differences and differences in care systems, their public, private or mixed funding, their administration systems, the role of families and the community" [17]. Felaibe's book, "Bioethics and Humanization of Clinical Care," addresses topics such as the empowerment of pediatric patients with chronic conditions, human rights involved in mental health care, death and dignity of the ill person, palliative care, death in vulnerable populations, elective cesarean sections requested by the patient, and gender dysphoria [19], demonstrating that clinical ethics problems can arise throughout the entire life cycle [18].

Decisions in medicine have both a scientific-technical and an ethical component; the best decisions will consider both. The scientific-technical component involves clinical aspects like diagnosis, prognosis, available treatments and access, and risk–benefit analysis of each therapy. The ethical component ensures that decisions made seek the greatest good for the person seeking care, requiring consideration of the patient's context, biography, quality of life, opinion, criteria, and preferences. For Carmen Astete, "the conception of a good clinical decision has not changed over time, but undoubtedly, what has changed is how it is reached," because in a clinical relationship model that promotes patient participation, "the greater good is defined based on the values, principles, beliefs, expectations, sense of life, illness, and death of the person affected by the decision" [20].

The complexity of decision-making requires the contribution of clinical ethics, which "refers to decision-making and reflection about the best choice(s) to make in the delivery of health care" [21]. Clinical Ethics Committees have been proposed as one mechanism to support professionals and users in complex health decision-making and can serve as a concrete field of action for clinical bioethics.

6.2.1 Clinical Ethics Committees in Latin America

Clinical Ethics Committees (CECs) can be defined as collegiate, interdisciplinary, deliberative and consultative structures created to analyze and advise on resolving ethical problems identified in healthcare. They contribute to improving care quality and protecting patient rights related to their health care [22, 23].

I agree with Omar Franca-Tarragó that it is difficult to accurately portray the institutionalization of CECs in Latin America. While an objective snapshot might be possible at a given moment, this information changes rapidly: new committees may form, some may not actively function, and others may exist on organizational charts but not genuinely perform their intended roles [24]. Furthermore, published experiences of these committees at the Latin American level are limited, and in certain countries, such as Peru or Panama, the development of CECs is less advanced compared to research ethics committees in the same region [25, 26].

Despite the challenges mentioned, it is worth noting that an increasing number of Clinical Ethics Committees are being progressively established in various countries. These committees are known by different names, such as Hospital Ethics Committees or Hospital Bioethics Committees, Healthcare Ethics Committess, among others. In certain contexts—such as in Chile—where they are called "Comités de Ética Asistencial" (which may be traslated as "Healthcare Ethics Committees"), these committees have been established not only in clinical settings (hospitals and primary care centers) but also within municipalities and health services. In Latin America, the first Hospital Ethics Committee was established in Argentina in 1984, followed by the Mainetti Foundation organizing the First Course on Hospital Ethics Committees in 1989 [27]. In other countries, Clinical Ethics Committees were formed in the 1990s, as in Chile [28] and Brazil, where "in 1993, the Hospital das Clínicas de Porto Alegre created the first Hospital Bioethics Committee (CBH), followed by the Hospital das Clínicas of the Faculty of Medicine of the University of São Paulo and the National Cancer Institute, in Rio de Janeiro" [29].

Additionally, various countries have norms or regulations that promote and govern the formation of these committees. For example, some of these regulations include:

- In Argentina, a 1993 ministerial resolution mandated self-managed hospitals to form hospital ethics committees. Subsequently, in 1996, Law 24.742 was enacted, stating that "in every hospital of the public health and social security system, to the extent that its complexity allows, there must be a Hospital Ethics Committee" [30, 31].
- In Chile and Uruguay, patient rights and duties laws require these committees in public and private hospitals. Specific regulations then govern their formation and operation in each country: Chile has Decree 62 of 2012 [23], and Uruguay currently has the Regulation for the Functioning of the Institutional Bioethics Commission [32, 33].

- In Colombia, Article 2 of Resolution 13437 of 1991 establishes the formation of Hospital Ethics Committees and the adoption of the Patients' Bill of Rights. Furthermore, Articles 15 and 16 of Decree 1757 of 1994 regulate their formation, functions, and operation [34, 35].
- In Ecuador, Ministerial Agreement-0005, "Expídese el Reglamento sustitutivo del reglamento para la aprobación y seguimiento de Comités de Ética de Investigación en seres humanos (CEISH) y Comités de Ética Asistenciales para la Salud (CEAS)" regulates their formation and operation of Healthcare Ethics Committes as well as case evaluation. It also mandates that Healthcare Ethics Committees must be approved by the Ministry of Public Health [36].
- In Mexico, "The Decree that adds Article 41 Bis and reforms Article 98 of the General Health Law—published in the Official Gazette of the Federation on December 14, 2011—introduces the obligation to have Hospital Bioethics Committees (CHB) in healthcare establishments of the public, social, or private sector of the national health system" [37].
- In Panama, Resolution No. 05673 of 1998 establishes "to create in each of the Public Hospitals, under the administration of the Ministry of Health, the Hospital Bioethics Committee" [38].

6.2.2 Functions of Clinical Ethics Committees

Clinical Ethics Committees are established as "spaces for reflection, analysis, guidance, and education" [37]. Through their functions, clinical ethics fulfills its objective of advising on decision-making and reflecting on the best options for each patient receiving care.

Clinical Ethics Committees typically have three key functions: consultative, educational, and normative. The consultative function involves providing advice on clinical cases that require ethical guidance. This function supports decision-making on complex issues by identifying and resolving problems that arise in the healthcare of patients of all ages. It formulates recommendations that aim to balance scientific-technical aspects, patient values, and ethical principles, while also explicitly stating their ethical foundations [27, 28]. As these recommendations are consultative, they are not binding and do not replace the responsibility of the physicians [37].

This consultative function can be carried out through committee meetings or by clinical ethics consultants. In some Latin American countries, such as Colombia, Brazil, and Chile, experiences with clinical ethics services or ethical consulting systems have been published. In these systems, individual consultants or small groups fulfill this role, bringing clinical ethics closer to the patient's bedside and complementing the functions of the Clinical Ethics Committee [39–42].

Various studies have identified ethical problems in Latin American countries related to respect for autonomy, clinical relationship, and to withhold or withdraw life-sustaining treatments (referred to as "appropriateness of therapeutic effort" in Latin America context) [10]. In primary care, a Chilean study highlighted ethical issues at this level of care, including problems related to access, resource allocation and management, lack of time and space for adequate care, confidentiality and information management, the balance between benefit and respect for autonomy, the role of parents in decision-making for minors, the impact of social determinants on health, drug use and lifestyle among adolescents, care for adults with chronic noncommunicable diseases, abandonment and infantilization of older adults, and end-of-life issues in both children and adults [43].

Publications mentioning the reasons for consultation to clinical ethics committees in Latin America refer to: appropriateness of therapeutic effort and analysis of therapeutic proportionality, competence and absence of patient representation, ethics of resource distribution, communication failures, families in denial about the patient's prognosis, truth-telling, refusal of treatments, voluntary discharge, decision-making related to maternity, interruption of pregnancy, requests from family members for treatments with questionable medical indications, transgender patients requesting sex reassignment surgery outside comprehensive care programs, problems related to confidentiality, rights of older adults, children, adolescents and migrants, among others [14, 44–46].

Ethical problems associated with end-of-life decisions are common across different realities. A terminally ill patient may be offered numerous care options, ranging from comfort-focused care to more complex and invasive management or procedures. "The person who in principle makes the decision is the patient, based on their right to accept or reject treatments" [47]. However, at the end of life, the patient's capacity to make decisions may be compromised. In such situations, it's essential to ascertain if the patient had previously expressed their wishes and criteria, either verbally or in advance directive documents [47]. Advance care planning and advance directive documents are seen as ways to promote and project autonomy for when a patient can no longer express their opinions. In Latin America, some countries have legislation on advance directives for healthcare: Puerto Rico, Panama, Argentina, Mexico, Uruguay, Colombia, Costa Rica, and more recently, Chile with the Universal Palliative Care Law [48]. Brazil still lacks specific legislation on the rights of terminally ill individuals, though some resolutions from the Federal Council of Medicine (CFM) address this issue [49].

The educational function refers to providing training in bioethics to CEC members, as well as to the professional and user community. This role is crucial, as CEC members need continuous training in bioethics and emerging bioethical issues relevant to their realities. This ensures that they contribute effectively to their institutions, improving care quality and promoting humanized care that considers the user's well-being [32].

Regarding the bioethics training of CEC members themselves, it is a relevant aspect because it "will allow for a more valuable and even more fruitful deliberation in terms of better substantiating the recommendations or regulations that emerge from this working group" [28]. Bioethics education in Latin American universities began in the 1980s, solidifying as an academic discipline in several regional countries in the 1990s. Notably, the Pan American Health Organization, in collaboration with the University of Chile, created the Regional Bioethics Program in 1994 [50], which is now managed by PAHO's Central Office in Washington. The consolidation of bioethics as a discipline in Latin America fostered training in this area at both undergraduate (medicine and health careers) and postgraduate levels. Currently, various continuing education and postgraduate bioethics programs exist in Latin American countries such as Mexico, Colombia, Argentina, and Chile, among others, promoting the acquisition of knowledge and competencies in this discipline. A concrete example of bioethics training for CEC members is offered by the Bioethics Office of the Ministry of Health of Chile, by organizing annual conferences and seminars, in line with one of its objectives being to collaborate on bioethics education. This office also manages the development of recommendation documents [51].

It's also important that this bioethics education extends throughout the entire healthcare community, reaching both healthcare workers and users. Many healthcare ethics committees, at least in Chile, organize bioethics conferences that invite professionals from their own institutions.

Through their consultative role and by promoting ethical training for institutional professionals, Clinical Ethics Committees, and where established, clinical ethics consultants or clinical ethics sevices, can also support in addressing the moral distress experienced by healthcare professionals [52].

Finally, the normative function refers to the CEC's contribution to institutional policies and protocols. This includes participating in their development and creating guidelines for recurring clinical ethics problems or those arising from the analysis of similar cases [28]. It can also involve participating in proposals aimed at improving care quality and modifying work methods within the institution [27].

6.2.3 Challenges of CECs

It's important to recognize that Clinical Ethics Committees must be attuned to their local realities and idiosyncrasies to truly contribute to their institutions. This is crucial in Latin America, where clinical ethics problems occur in specific contexts and situations, requiring sensitivity to these conditions.

For CECs to function effectively, they need institutional support that understands the importance of these collegiate deliberative bodies [32], which implies adequate structure, resources, and recognition [53]. Their institutional autonomy regarding the recommendations they issue and the continuous training and qualification of their members must be respected [32]. Committees should have a systematized work process and a multidisciplinary composition, with diversity in gender and age among members, to ensure that different views and opinions represent multiple moral convictions and enrich the deliberation [28]. Undoubtedly, a fundamental aspect is to increase awareness among both users and professionals within the institution about healthcare ethics committees, as this is the only way their advice and guidance will be sought [53].

Clinical Ethics Committees should be spaces that foster trust, where anyone identifying a clinical ethics problem feels welcomed and not judged [53]. These committees should possess the characteristics highlighted by Rogelio Altisent et al.: attuned to the real needs of professionals and users in their specific healthcare setting; accessible to anyone needing support, including professionals, users, and families; agile in their responses without compromising necessary deliberation, but providing responses or recommendations within a useful timeframe for the consultant; and maintaining efficiency and continuous evaluation [53]. This ongoing evaluation will allow identifying strengths, weaknesses, new needs, and contexts where they can contribute, thereby improving the service they aim to provide to the community [28].

6.3 Conclusion

In current medical practice in Latin America, identifying clinical ethics problems that generate complex challenges and decisions is increasingly possible. Therefore, the contribution of clinical ethics in decision-making becomes necessary. Deliberation and reflection are required, promoting shared decision-making with appropriate patient and family participation, considering their cultural, social, and value contexts. This demands greater professionalization of bioethics, with a progressive increase in the number and training of professionals to carry out healthcare work in this discipline, alongside the bioethics training of healthcare professionals.

Clinical Ethics Committees in various Latin American countries represent a concrete form of clinical bioethics development. Their consultative function, through deliberation on clinical ethics problems identified in different clinical encounters, allows them to support professionals, patients, and users. Their educational function trains the professional and user community in bioethics, and their normative function enables participation in the development of institutional protocols and policies. These committees must be attentive to the challenges they face, as well as to the diverse realities in which they are embedded, to provide patient-centered medicine.

References

1. Esquerda M. Retos éticos en Ética Clínica I: La bioética necesita apellidos. Blog de Bioética IBB; 2025. Available from: https://www.iborjabioetica.url.edu/es/blog-de-bioetica-debat/retos-eticos-en-etica-clinica-i-la-bioetica-necesita-apellidos.
2. Valera L. Bioética y bioética clínica. In: Beca JP, Astete C, Carvajal S, editors. Bioética Clínica. 2nd ed. Ed. Mediterráneo Ltda.; 2022. p. 19–24.
3. De los Ríos Uriarte M.E. La cuestión del método en Bioética clínica: aproximación a una metodología adaptada al contexto de la realidad mexicana. Pers Bioét. 2017;21(1):92–113.
4. Lolas SF. Bioética y contexto social. In: Bioética en América Latina Una década de evolución. Monogr Acta Bioeth, vol. 4. CIEB Universidad de Chile and the PAHO/WHO Bioethics Program; 2002.

5. Pessini L. Bioética: algunas interrogantes que desafían el presente y el futuro de América Latina. In: Ferrer JJ, Lecaros JA, Molins R, editors. Bioética: el pluralismo de la fundamentación. Universidad Pontificia Comillas; 2016. p. 419–32.
6. Rodríguez-Suárez N, Prieto-Martínez P. The impact of clinical ethics consultations on physicians in a Latin American context. Asian Bioeth Rev. 2024;16(4):635–51.
7. Álvarez-Díaz JA. ¿Bioética latinoamericana o bioética en Latinoamérica? Rev Latinoam Bioet. 2012;12(1):10–27.
8. Carvajal S. La relación clínica. In: Beca JP, Astete C, Carvajal S, editors. Bioética Clínica. 2nd ed. Ed. Mediterráneo Ltda.; 2022. p. 99–110.
9. Lazcano-Ponce E, Angeles-Llerenas A, Rodríguez-Valentín R, et al. Communication patterns in the doctor–patient relationship: evaluating determinants associated with low paternalism in Mexico. BMC Med Ethics. 2020;21:125. https://doi.org/10.1186/s12910-020-00566-3.
10. Blanco PA, García-Caballero R, Real de Asúa D, et al. What ethical conflicts do internists in Spain, México and Argentina encounter? An international cross-sectional observational study based on a self-administered survey. BMC Med Ethics. 2024;25:123. https://doi.org/10.1186/s12910-024-01123.
11. Ortiz A, Burdiles P. Informed consent. In: Beca JP, Astete C, Carvajal S, editors. Bioética Clínica. 2nd ed. Ed. Mediterráneo Ltda.; 2022. p. 123–33.
12. Garrafa V, Osório de Azambuja L. Epistemología de la bioética—enfoque latinoamericano. Rev Bras Bioética. 2007;3(3):344–59.
13. Garcia LF, Fernandes MS, Moreno JD, Goldim JR. Mapping bioethics in Latin America: history, theoretical models, and scientific output. J Bioeth Inq. 2019;16:323–31.
14. Nadal C, Campillay MC. Trayectoria de un comité de ética en la atención primaria. Rev Bioét. 2024;32:e3645PT.
15. Litewka S. Telemedicina: un desafío para América Latina. Acta Bioeth. 2005;11(2):127–32.
16. Esquerda M. Discurso de entrada en la Academia Médico-Quirúrgica Española: La ética médica como ética de la virtud: el legado de Pellegrino. 2025. Available from: https://es.linkedin.com/posts/montse-esquerda-02374452_discurso-de-entrada-en-la-amqemesquerdapdf-activity-7337123956174245888-LX-j.
17. Silva P. Prólogo. In: León FJ, editor. Bioética y humanización en la práctica clínica. Santiago: Fundación Interamericana Ciencia y Vida; 2017. p. 5–6. Available from: https://felaibe.org/material-descargas-articulos-felaibe.
18. Beca JP. La deliberación ante un problema ético-clínico. In: León Correa FJ, editor. Ética Clínica y Comités de Ética en Latinoamérica. FELAIBE; 2011. p. 155–62. Available from: https://felaibe.org/material-descargas-articulos-felaibe/.
19. FELAIBE. In: León FJ, editor. Bioética y humanización en la práctica clínica. Santiago: Fundación Interamericana Ciencia y Vida; 2017. Available from: https://felaibe.org/material-descargas-articulos-felaibe/.
20. Astete C. Voluntades anticipadas y planificación anticipada de decisiones. In: Beca JP, Astete C, Carvajal S, editors. Bioética Clínica. 2nd ed. Ed. Mediterráneo Ltda.; 2022. p. 135–43.
21. Moodley K. Bioethics: clinical. In: Ten Have H, editor. Encyclopedia of global bioethics. Cham: Springer; 2015. https://doi.org/10.1007/978-3-319-05544-2_45-1.
22. Couceiro A. Los comités de ética asistencial: origen, composición y métodos de trabajo. In: Couceiro A, editor. Bioética para clínicos. Madrid: Triacastela; 1999. p. 269–81.
23. Ministerio de Salud de Chile. Decreto 62. Aprueba reglamento para la constitución y funcionamiento de comités de ética asistencial. 2012. Available from: https://www.bcn.cl/leychile/navegar?idNorma=1052731
24. Franca-Tarragó O. Dificultades y Posibilidades de los Comités de Ética Desde una Perspectiva Latinoamericana. J Int Bioethique. 2007;18(1):147–53.
25. Shiraishi-Zapata C. The paucity of clinical ethics committees in Peru. Dev World Bioeth. 2022;22(4):185–6.

26. Comité Nacional de Bioética de la Investigación Panamá. Comités de Bioética Institucionales Acreditados. 2025. Available from: https://cnbi.senacyt.gob.pa/comites-acreditados/. Accessed 3 July 2025.
27. Gómez MI. Los Comités de Ética Asistencial. In: Reflexiones sobre Bioética: Seminarios de la Academia Chilena de Medicina (2011–2013). Academia Chilena de Medicina; 2015. p. 170–9.
28. Portales MB. Comités de ética asistencial y consultoría ético clínica. In: Beca JP, Astete C, Carvajal S, editors. Bioética Clínica. 2nd ed. Ed. Mediterráneo Ltda.; 2022. p. 145–511.
29. Sayago M, Amoretti R. Comités de bioética hospitalaria: importancia, funcionamiento y dificultades de implementación. Rev Bioét. 2021;29(4):831–42.
30. Sistema Público de Salud de Argentina. Ley 24.742. Comité Hospitalario de Ética. Funciones Integración; 1996. Available from: https://www.argentina.gob.ar/normativa/nacional/ley-24742-40993/texto.
31. Rabadán AT, Tripodoro VA. ¿Cuándo acudir al comité de bioética institucional?: El método deliberativo para resolver posibles dilemas. Medicina (B Aires). 2017;77(6):486–90.
32. Heus Texeira LC. Los Comités de Bioética Asistencial. In: Colegio Médico del Uruguay. Temas de Bioética; 2021. p. 75–87.
33. Ministerio de Salud Pública de Uruguay. Ordenanza N° 1.588/020. Reglamento de Funcionamiento de la Comisión de Bioética Institucional; 2019. Available from: https://www.gub.uy/ministerio-salud-publica/institucional/normativa/ordenanza-n-1588020-reglamento-funcionamiento-comision-bioetica.
34. Ministerio de Salud y Protección Social de Colombia. Resolución 13437. 1991. Available from: https://www.minsalud.gov.co/Normatividad_Nuevo/RESOLUCIÓN%2013437%20DE%201991.pdf.
35. Ministerio de Salud y Protección Social de Colombia. Decreto 1757. 1994. Available from: https://www.funcionpublica.gov.co/eva/gestornormativo/norma.php?i=21850.
36. Ministerio de Salud Pública de República del Ecuador. Acuerdo N° 00005-2022. 2022. Available from: https://www.salud.gob.ec/wp-content/uploads/2022/09/A.M.-00005-2022-JUL-29.-QUINTO-SUPLEMENTO-NO.-118-SUSTITUTORIO-4889_compressed.pdf.
37. Comisión Nacional de Bioética. In: Ruiz de Chávez M, editor. Guía nacional para la integración y el funcionamiento de los Comités Hospitalarios de Bioética. México: Secretaría de Salud; 2015. Available from: https://www.conbioetica-mexico.salud.gob.mx/descargas/pdf/registrocomites/Guia_CHB_Final_Paginada_con_forros.pdf.
38. Ministerio de Salud de Panamá. Resuelto No. 05673. Gaceta Oficial; 1998. Available from: https://www.gacetaoficial.gob.pa/gacetas/23663_1998.pdf.
39. Gempeler FE, Torregrosa L, Silva JM. Clinical bioethics by the patient's side. "Development and experience of a clinical ethics service" in a university hospital. Univ Méd. 2017;58(2):1–7.
40. Prieto-Martínez P, Rodríguez SN, Zárate-Velasco DL. Department of Humanism and Bioethics at the Fundación Santa Fe de Bogotá, Colombia: an update based on 8 years of service. Clin Ethics. 2023;18(4):393–403.
41. Melnik CS, Goldim JR. Profile of clinical bioethics consultancies involving families which hindered the resolution of problems. Rev Bioét. 2013;21(1):108–13.
42. Beca JP, Koppmann A, Chávez P, Delgado I, Solar S. Análisis de una experiencia de consultoría ético clínica en cuidado intensivo. Rev Med Chile. 2010;138(7):815–20. https://doi.org/10.4067/S0034-98872010000700004.
43. Contreras AL. Problemas ético clínicos en la Atención Primaria del Centro de Salud Familiar de Paine. Acta Bioeth. 2017;23(1):25–34.
44. Vega S, Parry S, Pérez G, et al. Experiencia del Comité de Ética Asistencial del Hospital Carlos van Buren entre los años 2007–2020. Rev Med Chile. 2022;150(3):391–6.
45. Weingerz MS, Templos Esteban LA, Rangel Domínguez NE, Orellana Caro VR. Motivos de presentación de casos clínicos de pacientes ante el Comité Hospitalario de Bioética, en un Hospital de Segundo Nivel. Med Etica. 2021;32(3):703–57.

46. Valenzuela S. Comité de Ética Asistencial hospital Clínico de la Universidad de Chile: memoria de 10 años de trabajo. Rev Hosp Clín Univ Chile. 2018;29(3):223–8.
47. Astete C, Beca JP. Decisiones clínicas al final de la vida. In: Beca JP, Astete C, Carvajal S, editors. Bioética Clínica. 2nd ed. Ed. Mediterráneo Ltda.; 2022. p. 243–52.
48. Monteiro RSF, Silva AGD. Directivas anticipadas de voluntad: recorrido histórico en América Latina. Rev Bioética. 2019;27:86–97.
49. Borges LS, Lima MJV. Voluntades anticipadas y cuidados paliativos: percepción brasileña. Rev Bioét. 2024;32:e3636PT.
50. León FJ. Fuentes para una historia de la bioética en Latinoamérica y su desarrollo. In: La bioética latinoamericana en sus textos. Centro de Estudios Interdisciplinarios en Bioética (CEIB); 2008.
51. Oficina de Bioética del Ministerio de Salud de Chile. 2025. Available from: https://bioetica.minsal.cl/.
52. Cambra FJ. Distrés moral. In: Congreso de Bioética. Hospital San Juan de Dios; 2025. Available from: https://www.laborhospitalaria.com/wp-content/uploads/2025/01/LH_340_02_6_congreso.pdf.
53. Altisent R, Fernández-Letamendi T, Delgado-Marroquín MT. Una nueva vitalidad para el futuro de los comités de ética asistencial. Folia Humaníst. 2019;13:19–33.

Part III
Global Bioethics

Chapter 7
Demographic Winter: A Global Comparative Analysis of Regional Trends, Socioeconomic Impacts, and Policy Imperatives

Carlo De La Llata-Gómez and Luis Medina-Gual

7.1 Introduction

7.1.1 Definition of Demographic Winter

This chapter addresses the phenomenon of demographic winter, characterized by low fertility rates that result in insufficient generational replacement, population aging, and the potential reduction of its total size [1]. This process, an advanced phase of the universal and ineluctable demographic transition [2], initially reduces infant mortality [3] and fertility. It implies significant changes in age structures [4] and presents major economic and social consequences at a global level [2, 4].

Its global relevance lies in the challenges for social security systems, healthcare, and labor markets, given the larger proportion of elderly dependents and the smaller proportion of working-age people [1, 3–5]. These factors alter consumption and savings dynamics and project a significant increase in health and pension costs if current patterns are maintained [4].

C. De La Llata-Gómez (✉)
Universidad Anáhuac, Mexico City, Mexico

L. Medina-Gual
Universidad Iberoamericana, Mexico City, Mexico
e-mail: luis.gual@ibero.mx

D. Cerdio-Domínguez, M. E. de los Ríos-Uriarte (eds.), *Current Bioethics*, New Paradigms in Healthcare, https://doi.org/10.1007/978-3-032-25294-4_7

7.1.2 Justification for the Comparative Regional Approach

The manifestation of demographic winter varies notably. Western Europe, with an early onset and advanced progression of the phenomenon, faces acute challenges of generational replacement and altered age structures, where immigration has partially compensated for low fertility [1]. In developing regions, the situation is heterogeneous [2]. The Arab world shows rapid but disparate sociodemographic and health transitions, with diversification of demographic systems and internal inequalities; the changes have been radical in the Maghreb and Iran, slower in Saudi Arabia or Egypt, and marginal in Yemen [6]. Some studies have indicated that the educational level, especially maternal, is key to reproductive behaviors in a context with still considerable illiteracy rates in some countries (32% among adults around 2002 in the Arab region) [6].

On the other hand, Latin America and the Caribbean (LAC) presents multiple demographic transitions, with countries that are aging (e.g., Argentina, Cuba, and Uruguay) and others with strong growth and high fertility rates (e.g., Bolivia and Guatemala) [7]. Aging in LAC will occur faster than in developed countries [3]. However, many of these countries are still experiencing the "demographic dividend," a window of opportunity to invest in education and prepare for future aging [4, 7].

In contrast, Sub-Saharan Africa, although showing a decreasing trend in fertility [8, 9], maintains higher rates and a slower transition compared to other developing regions [8, 9]. Factors such as an improvement in economic living standards, a reduction of infant mortality, structural-economic changes (such as the reduction in the proportion of nonagricultural labor in the face of growing industrialization), the increase in the basic educational level, particularly in the female sector, and the formalization of economic activities are crucial for the decline in fertility in this region [8].

This heterogeneity justifies a comparative regional approach to understanding the interactions between socioeconomic, cultural, educational, and political factors in demographic trends and their consequences [6–8]. This chapter will explore these dynamics, seeking to offer a nuanced understanding of the interrelationships between demography and global socioeconomic development by examining regions such as the Arab world [6, 10], Sub-Saharan Africa [8, 9], LAC [4, 7], and Europe [1]. Thus, this chapter will be based on a literature analysis conducted by modifying the PRISMA methodology [11] using various specialized artificial intelligence tools, including Elicit and Scopus AI.

7.2 Analysis by Region

This section analyzes each contemplated region—Arab Africa and the Middle East, Western Europe, Eastern Europe, Latin America, North America, and non-Arab Africa—organizing the literature analysis around common elements. First, each region is contextualized from a demographic perspective, addressing aspects such as fertility and birth rates, whether high or low, and their historical and structural causes. Next, the factors that, according to the reviewed literature, influence the phenomenon of demographic winter are identified and described. These factors include economic, social, political, cultural, and religious dimensions. Finally, it concludes with a synthesis of the consequences observed in the region, an analysis of the public policies implemented, recommendations formulated in the literature, and possible lines of action to face the challenges associated with demographic transformations.

7.2.1 *Arab Africa and the Middle East*

7.2.1.1 Recent Demographic Trends (Decline in Fertility, Increase in Delayed Marriage, and Use of Contraception)

The Arab world region is immersed in a rapid demographic transition [12], characterized by substantial transformations in fertility, mortality, and migration profiles that began to take shape at the end of the twentieth century [6, 12]. This process started with a marked reduction in infant mortality [10]; however, this decline occurred with a certain lag relative to other regions [6]. This sequence of events resulted in a notable "demographic surplus," the direct consequence of which was the high rates of natural growth observed during the period [10].

The central feature that followed was the reduction in fertility, initiated in the late 1970s [10, 13], a phenomenon that marked the end of the so-called "demographic explosion" [10]. However, this dynamic exhibits considerable regional heterogeneity [6, 10] and, in some countries, recent upticks [14, 15]. Concurrently, other demographic trends have been accentuated, including the increase in delayed marriage and the emergence of permanent celibacy [6, 10, 13, 16]; the expansion of contraceptive use [16–18]; continuous urbanization—predominantly coastal [6, 10, 15]; and significant international migration flows. The change in the age structure, by increasing the proportion of people of working age relative to the dependent population, allowed for favorable conditions for the so-called "demographic dividend" [6, 19], which is a phenomenon characterized by the boost in economic growth that a country can experience when a large part of its population joins the labor market, raising productivity and savings, provided that there are adequate policies that favor it; however, the persistence of illiteracy in these countries still constitutes a structural obstacle [6, 10].

7.2.1.2 Structural Factors (Female Education, Urbanization, Migrations, and Regional Inequalities)

To understand the demographic transition in the Arab region, it is essential to analyze how the various structural factors, closely interrelated, are organized around the idea of "openness" to external influences [10]. The reduction in fertility is due, in the first instance, to transformations in the proximate determinants—delayed marriage, birth control and family planning, abortion, and breastfeeding practices [6, 13, 16, 17, 20]. However, this decline is also within a framework of indirect or underlying determinants. Among these, education stands out, especially female education—a crucial factor whose expansion was, despite its importance, late and incomplete [6, 10, 13, 21]; massive urbanization associated with the rural exodus, a phenomenon inseparable from lower fertility levels [10, 21]; internal and international migrations, which reshape reproductive behaviors; the status of women and their incorporation into the labor market [10, 21]; and various economic variables that affect family decisions [18, 21]. Finally, the political, cultural, and religious background fundamentally impacts demographic dynamics, as population policies, influenced by these factors, largely determine birth, migration, and mortality rates. The tension between "openness" and "closure," which refers to the receptiveness or restriction of migration and reproductive policies, reflects how political, cultural, and religious decisions affect the structure and growth of a country's population [10, 13].

7.2.1.3 Socioeconomic Consequences and Challenges for Development (Pressure on Resources, Intra-Country Differences, and Unequal Access to Services)

Demographic evolution has had profound socioeconomic repercussions [10, 15], reflected in the family structure, the economy, the labor market, education, and health [6, 10]. Population growth demands a costly "demographic investment" in infrastructure and services [10]. Among the central challenges are managing both development and the transition [12], financing this investment, reversing the "agricultural failure" [10], and reducing youth unemployment [10]. Added to this are the sociocultural tensions between modernity and tradition—the "pesanteurs" or forces of "closure"—the interference of political regimes [10], and the scarcity of reliable demographic data [10, 16]. In this context, the status of women emerges as a key factor [10].

7.2.1.4 Public Policies (Needs for Educational, Health, and Inequality Reduction Policies)

The demographic challenges consequently demand a comprehensive and multifaceted policy response [6, 10]. In the educational sphere, it is imperative to universalize schooling—particularly for females—and raise the teaching quality [10]. In terms of health, it is essential to deepen the reduction of mortality through the expansion of reproductive health and effective access to family planning and contraceptive methods; all articulated with actions that strengthen the status of women [6, 10, 15]. From the economic sphere, policies must confront agricultural failure, boost investment, and advance in the liberalization of markets [6, 10]. Concurrently, it is necessary to foster a political, economic, and cultural "openness"—which includes freedom of the press and democratization—and to promote a form of governance capable of reconfiguring the relationship between the state and social forces [10]. Ultimately, the political–economic decisions of governments are often more determinant than cultural or religious factors considered in isolation [10].

7.2.2 Western Europe

7.2.2.1 Recent Demographic Trends (Low Fertility, Accelerated Aging, and Late Age at Motherhood)

Since the 1970s, Western Europe has undergone a profound demographic transformation, marked by a sharp decline in fertility and rapid population aging [1, 22]. The Total Fertility Rate (TFR) declined from the peak levels of the 1960s [1]; thus, Italy went from 2.70 children per woman in 1964 to 1.19 in 1995, while comparable drops were recorded in Ireland and Portugal [1]. Since the 1990s, almost no European country has maintained fertility above the replacement threshold (2.1 children per woman) except for temporary episodes [1]. In the south of the continent—with Italy at the forefront—fertility is approaching "very low" values (< 1.3), a threshold already reached by a third of the countries in 2002 [1]. This is part of the so-called Second Demographic Transition (SDT), which entails the postponement of motherhood and the first child, increasing extramarital births, reducing higher birth orders, and expanding permanent childlessness [1, 23]. The 2008–2011 recession deepened the birth decline, especially in Southern Europe and among the young population [1].

In parallel, population aging has intensified: Europe has the highest median age in the world (41.6 years) [1], with highs of 45.9 years in Italy and Germany [1]. Initially driven by low fertility ("aging from the bottom"), this process has been amplified by the reduction of mortality at advanced ages ("aging from the top") and the consequent extension of life expectancy [1, 22]. The current age pyramid conditions the immediate future, as the scarcity of women of reproductive age makes a low birth rate foreseeable in the short and medium terms [1]. In this context, immigration emerges as the primary source of demographic growth, especially in countries with negative natural balances, such as Germany, Greece, and Italy [1, 22]; its contribution is manifested directly and through the births of the immigrants and their descendants [1, 24].

7.2.2.2 Socioeconomic and Cultural Factors (2008 Economic Crisis, Changing Values, and Work–Family Reconciliation Policies)

As argued, low fertility rates and accelerated population aging result from the convergence of economic factors, social transformations, political decisions, and cultural shifts [1, 25, 26]. Within the framework of the Second Demographic Transition, a contraction of nuptiality and fertility, the rise of cohabitation and extramarital births, and the postponement of motherhood and fatherhood are observed in European countries [1, 26, 27]. These behaviors are related to secularization, the acceptance of family diversity, and the massive incorporation of women into education and the labor market [1, 26, 28], weakening the traditional link between reproduction and marriage [1, 22]. Therefore, the reconciliation of the family and professional spheres emerges as a central challenge [22, 28, 29]. At the same time, in Southern Europe, the younger generations show a clear tendency to forgo having a first or second child, reflecting a shift in demographic and social priorities [1].

The economic dimension is decisive [26, 28]. The global economic crisis of 2008 significantly affected Europe, especially the countries of the southern continent. The economic crisis caused a deep recession, generating high unemployment, job precarity, and economic instability. This adverse context contributed to a reduction in the birth rate, particularly among young people [1], as economic conditions and the lack of institutional support (such as welfare policies and child-rearing subsidies) caused many people to delay or abandon their plans to have children. Economic uncertainty and the high cost of child-rearing were key factors in this trend [1, 28].

7.2.2.3 Consequences of Demographic Winter (Declining Labor Force, Pressure on Pension Systems, and Migratory Challenges)

The "demographic winter" in Western Europe, characterized by low birth rates, population aging, and, in some cases, demographic decline [1, 22], has significant economic and social repercussions. Economically, it generates pressure on pension, health, and social security systems due to the intergenerational imbalance and the growing dependency of the older population [1, 22, 27, 28]. Furthermore, it reduces labor supply, diminishes productivity and innovation capacity, and limits growth potential [1]. Socially, the imbalance in the age pyramid increases the risk of a self-sustaining population decline [1]. On the other hand, transformations are observed in family structure and reproductive patterns, such as greater childlessness, the expansion of "zero or one-child" family models, and a growing gap between the desired number of children and the number had, especially in Southern Europe [1, 22, 28, 29]. Consequently, an increasingly aging population demands reconfiguring social, health, and care services for older adults [22, 29].

7.2.2.4 Public Policies and Strategies (Promotion of the Birth Rate, Immigration Management, and Integrated Policies)

In the face of "demographic winter," European public action rests on two main levers: promoting the birth rate and managing immigration [1]. Pronatalist measures operate in the medium and long terms [1, 22, 28, 29], focusing on facilitating the reconciliation of family and work life. However, the absence of a clear and active demographic strategy in the European Union (EU) that supports the birth rate constitutes an alarming deficiency [22, 28]. Where robust "family-friendly" policies have been deployed—for example, in France and the Nordic countries—the emphasis is on childcare services and support for maternal employment; in contrast, approaches limited to monetary incentives—as is the case in much of Southern Europe—show less effectiveness [1]. Immigration, in contrast, offers more immediate impacts by incorporating a young and active population [1, 24]; it covers labor shortages, boosts births, and compensates for negative natural balances [1]. However, substantive rejuvenation requires large migratory flows and effective integration processes, given the convergence of immigrant fertility with the native population [24, 30]. Consequently, both levers must be synergized, as neither can reverse the demographic decline [1]. The lack of a clear and well-founded political strategy that considers the situation's complexity and the tendency to focus only on simplistic solutions, such as isolated economic incentives, could worsen the problem rather than resolving it [1, 22].

7.2.3 Eastern Europe

7.2.3.1 Recent Demographic Trends (Population Decline, Aging, and Impact of Political Transition)

Following the collapse of communism in 1989, Central and Eastern Europe (CEE) and the former Soviet Union (FSU) experienced profound political, social, and economic transformations that altered their demographic dynamics [31, 32]. Since the 1990s, the birth rate plummeted—15 states reached levels of "lowest-low fertility" (1.1–1.3 children per woman) by 2004 [31, 33–35]—while mortality rebounded due to economic crises, deteriorating health services, and impoverishment, with notable peaks in Russia and Ukraine [32, 33, 36]. This was compounded by a massive emigration of young people, especially intense in Ukraine [32, 36, 37]. The combination of low fertility, high mortality, and intensified migration accelerated aging and depopulation: Ukraine lost almost 10 million inhabitants between 1993 and 2020 [32, 36]. This situation is likely linked to high unemployment rates, a fall in the standard of living, and precarious conditions for young families [33, 36]. The debate persists as to whether the collapse of the birth rate prolongs previous trends or derives directly from the postcommunist crisis: the Second Demographic Transition points to the diffusion of Western values, while the economic approach underscores insecurity and anomie as drivers of the postponement or reduction of fertility [31, 35, 38]. The result is a landscape of accelerated aging and depopulation, aggravated in Ukraine by the war that began in 2022 [36].

7.2.3.2 Determining Factors (Impact of Economic Crises and Changes in Marriage and Birth Patterns)

The post-1989 transformation in CEE and FSU triggered a persistent population decline and accelerated aging. This process is characterized by three main features [31]: (i) extraordinarily low birth rates—below 1.5 and even 1.3 children per woman [32, 34, 35, 39]; (ii) unfavorable mortality trends [31–33, 35, 39], with extreme examples such as Ukrainian life expectancy, which is lower than the world average [40]; and (iii) consistently negative net migration [32, 33, 35, 36]. Economic factors are decisive [33]: socioeconomic deterioration and successive crises are the immediate causes of fertility collapse [34, 35]. Uncertainty, job and income insecurity, persistent unemployment, and anomie encourage the postponement of procreation [31, 33, 34], while the recession reduced incomes and expanded poverty [33]. Concurrently, social and ideational transformations were recorded [35], such as the diffusion of Western values linked to the Second Demographic Transition [31, 33, 35] and the transitional anomie reflected in the increase in divorce and extramarital unions [35], factors that reinforced the negative demographic trajectory.

7.2.3.3 Consequences of Population Decline (Loss of Active Population and Pressure on Services)

The demographic decline entails profound economic and social repercussions [32, 33]. On the financial level, low fertility and adverse conditions reduce the labor force [34, 35]. In contrast, the massive migration of the young population in search of better living conditions further diminishes the Economically Active Population (EAP) index of the countries [33, 36]; in parallel, impoverishment, loss of income, and the disappearance of free services deteriorate the standard of living [33, 39]. From a social perspective, population aging is exacerbated [32, 36], and smaller households and families are observed [33, 38, 39], as are material difficulties for young people, and fewer marriages and more divorces [36]. Poverty and inequality restrict access to basic services [33, 39], and the decline in medical quality negatively impacts demographic indicators [36]. External factors, such as COVID-19 and the war in Ukraine, have intensified these regional adverse trends [36, 40].

7.2.3.4 Public Policies: Challenges and Responses in a Context of Transformation (Need to Develop Concrete Policies and Experiences of Success/Failure)

The negative demographic dynamic is a matter of grave concern [40]. Following the postcommunist transition, various countries are recording rapid aging and a decrease in population [32, 36], which makes it indispensable to design concrete and practical policies [31]. Although low fertility is firmly established, the debate continues over its causes—socioeconomic or other—and the socioeconomic explanation, despite its popularity, sometimes has limited scientific support [33, 35, 39]. The case of Ukraine described as a "demographic catastrophe" due to the war reinforces the need for a comprehensive recovery strategy that includes financial support for young families, increased birth payments, social protection for single-parent families and those with disabled children, restructuring of elderly care, health promotion, rural revitalization, and strengthening of the healthcare system [36]. In Lithuania, family policy began late, is still under development, and lacks a coherent interconnected system [33, 39]. Understanding the underlying factors is thus crucial for guiding effective demographic interventions [31, 33, 35].

7.2.4 Latin America

In Latin America and the Caribbean (LAC), demographic dynamics have changed substantially over the last century [4]. Far from a uniform transition, the region presents multiple trajectories, generating a marked heterogeneity among countries [2, 3, 7], a key aspect for interpreting its demographic challenges and opportunities [5]. The transition is reflected in the drastic fall in infant mortality, the increase in life expectancy, and the pronounced reduction in fertility [4].

7.2.4.1 Recent Demographic Trends (Advanced Demographic Transition, Incipient Aging, and Intercountry Inequalities)

The diversity of demographic trajectories in LAC is reflected in the countries' different transitional phases [5]. In the advanced stage are Argentina, southern Brazil, Chile, Cuba, and Uruguay, whose early declines in fertility moderated in the mid-twentieth century; today, they exhibit low population growth and very marked aging, with Cuba as an extreme case [2, 4, 7]. In contrast, in countries like Bolivia, Guatemala, and Haiti, which are still in intermediate or initial phases, the reductions in mortality and fertility began later, retaining relatively high birth and demographic growth rates [4, 5, 7].

The "demographic dividend" designates the window in which the proportion of the working-age population exceeds that of the dependent population; its duration is limited and varies among countries. It is anticipated that from 2030, the dependency ratio will increase again with the advance of population aging [2, 7].

The demographic heterogeneity of LAC results from the convergence of historical, transitional, and socioeconomic factors. First, the different population trajectories mark differences: countries with intense European immigration today present more aged populations. At the same time, those with a significant indigenous presence delayed the fall in mortality and fertility [4, 7]. Furthermore, these nations are in disparate phases of the demographic transition; some already present low fertility rates, while others retain high fertility [2–4, 7]. Since the mid-twentieth century, the implementation of family planning policies and programs—supported by governments and private institutions—accelerated the general decline in fertility, albeit with diverse approaches [7]. However, in the regions of almost all these countries, heterogeneous situations and marked internal gaps persist: the less-educated and low-income sectors tend to maintain more accelerated demographic growth, with higher birth rates [2, 7, 41]. Thus, the demographic variability in LAC reflects the interaction between its population history, the pace of the transition, and birth control policies [7].

7.2.4.2 Key Transition Factors (Education, Urbanization, Family Planning Policies, and Migration)

In the last five decades, LAC transitioned from a demographic explosion to a sharp decline in mortality and fertility faster than developed countries [3, 4, 7]. This change responds, above all, to three factors: (i) the processes of educational expansion and democratization—affecting, especially, the female sector—which tends to modify reproductive decisions, reduces future fiscal pressure, and prepares states for population aging [4, 42]; (ii) urbanization processes, which are linked to growing internal migration and the establishment of productive and industrial centers in specific geographic poles, generating high urban concentrations, which historically is associated with lower fertility rates [4, 41]; and (iii) family planning policies, promoted since the International Conference on Population and Development in Bucharest in 1974 and later redefined at the Cairo Conference in 1994, which expanded the use of modern contraceptive methods [2, 7]. Nevertheless, despite the general decline in fertility, broad socioeconomic gaps persist. In some countries, adolescent fertility is even on the rise, demonstrating the limited scope of these policies in reducing inequalities [7].

7.2.4.3 Socioeconomic Consequences (Sustainability of Pension Systems and Public Spending on Health and Education)

In recent decades, LAC has shifted from high-fertility rural societies to predominantly urban ones with progressively aging populations [4]. Over the next 50 years, this rapid aging—driven by a sustained decline in mortality and fertility—will substantially raise public expenditures on pensions and healthcare [4]. As Daniel Cotlear's World Bank volume on population aging in Latin America underscores, this shift is poised to reshape the composition of social spending and the institutional agenda.

That demographic turn brings socioeconomic and fiscal pressures requiring comprehensive public policy responses [4]. Socially, the labor exclusion of older persons deepens poverty across the life course and increases demand for minimum pensions and healthcare amid the growing burden of noncommunicable diseases; while such pensions mitigate poverty, they do not eliminate it, and the equitable distribution of inter- and intra-family transfers remains unresolved [4].

Fiscally, the new age structure will ease educational spending—by 1%–2% of GDP over four decades—yet it will accelerate pension and health costs; retirement outlays are projected to rise to levels comparable to those of the European Union (≈2.4% of GDP), with an even greater impact on healthcare. Indeed, policy choices may shape future spending more than demography itself [4].

For policymakers, the task is to design institutions suited to an aging society, balancing the needs of older adults with investment in youth and vulnerable sectors [4]. Leveraging the current demographic dividend to finance these investments, integrating population dynamics into development strategies, and addressing persistent inequality—not overcome by previous demographic policies—are long-term imperatives [4, 7].

7.2.4.4 Evolving Public Policies (Pension Reforms, Investment in Youth, and Support Programs for Older Adults)

Faced with the rapid aging that follows the demographic dividend in LAC [2, 7], governments must fully integrate the population dimension into their development strategies to mitigate pressures on social protection systems [2, 7]. Policy priorities are concentrated on: (i) reforming pension systems to ensure sustainability and expand coverage, complementing them with social pensions that reduce poverty in old age [2, 4]; (ii) strengthening health services in the face of the rise of noncommunicable diseases, ensuring their financial viability [4]; (iii) combining public and intra-family transfers to mitigate the poverty of older adults [4]; (iv) facilitating flexible labor transitions and prolonging the active life of those who lack a pension [4]; (v) incorporating the impact of policy decisions into the projection of public spending and long-term fiscal management [4]; and (vi) reorienting educational savings derived from lower enrollment toward financing pensions and health, taking advantage of the demographic dividend to invest in future generations [4]. These actions articulate a multidimensional response to the challenges of aging.

7.2.5 *North America*

7.2.5.1 Recent Demographic Trends. (Low Fertility, High Immigration, Population Aging, and Internal Redistribution)

Since the 1980s, the demographic dynamics of Canada and the United States have exhibited a marked dynamism characterized by four interrelated trends. First, both nations maintain historically low fertility rates: the U.S. Total Fertility Rate is near the replacement level, while Canada's is around 1.5 children per woman; a decline that is related to the postponement of motherhood as a consequence of the prolongation of years of study, the development of professional careers, and the search for economic stability by the young population [43–47]. Second, immigration compensates for weak natural growth. Although the United States receives the most significant absolute number of immigrants, Canada has a relative proportion of almost double, with one in five residents born abroad compared to one in eight in the United States [43, 44]. Third, internal redistribution is shifting the population's "center of gravity" toward the West and south of the United States and British Columbia and Ontario in Canada, shaping regional patterns of expansion and stagnation [43, 44]. Fourth, aging is accelerating the proportion of people aged 65 and over, accompanied by family transformations—more diverse households and extramarital births—and growing geographic and social inequalities visible in health and fertility [43,

44]. Collectively, these trends portend slower population growth and pose the main challenge of addressing territorial and socioeconomic divergences [43].

7.2.5.2 Structural Factors. (Influence of Education, Urbanization Patterns, and Role of International Migration)

The demographic dynamics of Canada and the United States are driven by education, urbanization, and migration [43]. First, education—especially for women—postpones motherhood and reduces the number of children [45, 48]; even "shadow education" (private tutoring) affects the classic trade-off between the quantity and quality of offspring [49]. These changes, in turn, challenge the capacity of educational systems to adapt [44]. Second, the population has shifted from concentrating in large metropolitan areas to moving toward smaller communities since the 1970s—a flow motivated by economic and quality-of-life reasons that, in the United States, explains the move from the "snowbelt" to the "sunbelt" [43, 44]. Finally, international migration is essential: in the United States, it contributes almost half of the population growth, compensates for low birth rates, and diversifies the ethnic composition [44, 50]. Immigration policies have evolved to open up to multiple origins [43, 44], and all indications are that immigration will continue, demanding greater attention to the integration and adaptation of newcomers [49].

7.2.5.3 Consequences and Challenges (Socioeconomic Impact, Sustainability of Welfare Systems, and Integration and Management of Inequalities)

Recent demographic changes—low fertility (more pronounced in Canada), higher life expectancy (also greater in Canada), accelerated aging, decisive international migration, and family transformations—are reshaping the socioeconomic foundations of Canada and the United States [43, 51]. This combination reduces the working-age population, slows economic growth, and strains pension systems and elderly care; immigration mitigates these pressures, although it does not eliminate them [44, 49]. The increase in longevity raises healthcare demand, while cohort fluctuations impact educational planning [44]. Furthermore, the contribution of migration diversifies the ethnic composition and poses challenges for integration and social cohesion [43].

Consequently, the main challenges consist of curbing the widening of territorial and social inequalities, ensuring the sustainability of welfare, adapting labor markets, and promoting the full integration of immigrants, all while simultaneously managing low fertility and the repercussions of aging [43, 44, 48, 49].

7.2.5.4 Public Policies (Responses to Low Fertility and Immigration, Management of Inequalities, and Adaptation of Public Services)

Public policies must anticipate demographic challenges through coordinated and multidimensional strategies [43, 44]. Low fertility—linked to child-rearing costs, wage structures, and educational levels—can be addressed through direct economic support from the state to households, wage equity policies, and educational improvements [45, 49]. The internal redistribution of the population requires regional development and urban planning programs that moderate territorial disparities [44]. Likewise, immigration, which is decisive for growth, requires the development of programs that facilitate their full social and economic integration [43, 52]. Since demographic changes affect education, health, social security, and even the national security sector and the armed forces and defense, the different sectors must adapt their policies and services to the new population scenarios [44]. From the public policy perspective, it is fundamental that these governments adopt a comprehensive approach that not only mitigates the effects of demographic change but also capitalizes on its opportunities. This implies articulating intersectoral strategies that promote equity, strengthen social cohesion, and foster balanced territorial development, prioritizing investment in families, the integration of migrants, and the structural adaptation of each state sector.

7.2.6 Non-Arab Africa (Sub-Saharan Africa)

7.2.6.1 Recent Demographic Trends (High Fertility Rates and Onset of Decline in some Contexts)

On a global scale, the Total Fertility Rate (TFR) stands at 2.53, close to the replacement level, and 91% of the population resides in countries with low or intermediate fertility [53]. Sub-Saharan Africa, in contrast, maintains a TFR of 5.1 and is undergoing a slower transition, initiated in almost all its states but with very uneven progress [53]. Intraregional disparities are marked: Southern Africa records 2.6 children per woman and is experiencing a "silent revolution" in fertility despite the impact of AIDS [53, 54], while the eastern, western, and central subregions double that figure, with stagnant cases (Benin, Zambia, and Nigeria) and others of rapid decline (Ethiopia, Rwanda, and Namibia) [53]. Ethiopia stands out for the wide rural–urban gap—Addis Ababa is already below replacement—and for lags in maternal health, female literacy, and rural food security [55].

7.2.6.2 Conditioning Factors (Reproductive Health, Access to Contraception, and Female Education)

Socioeconomic development catalyzes the reproductive transition, although its effect is not homogeneous and depends on the context [53]. Among the most influential drivers are the reduction of infant and child mortality and the expansion of female education, especially at the secondary and tertiary levels [53, 56–61]. The diffusion of contraceptives and the strength of family planning programs modulate the speed of fertility decline [8, 57, 58, 62]. Added to this is the structural transformation of the economy—growth of nonagricultural employment, industrialization, incorporation of highly educated women, and labor formalization—capable of reducing fertility by an average of 1–1.6 children [8, 59]. Additional factors include reproductive preferences and sociocultural norms, the degree of urbanization and household wealth, modernization frameworks, cultural roots, and intergenerational resource flows [8, 9, 53, 57]. HIV/AIDS introduces direct (morbidity–mortality) and indirect effects (stigma and variation in the demand for family planning) of diverse magnitude [53, 63]. Equally relevant are adolescent fertility and infertility associated with poverty and stigmatization [53, 63, 64]. In the rural areas of Ethiopia, migration, mobility toward nonagricultural sectors, the delay of marriage, and the adoption of smaller family ideals have favored the decline in fertility [8, 55].

7.2.6.3 Public Policies for Development and Harnessing the Demographic Dividend (Challenges of Demographic Transition, Risks, and Opportunities of the Demographic Dividend)

The demographic transition in Sub-Saharan Africa is distinguished by its complexity and heterogeneity, with direct effects on development [8, 9, 53, 55]. The decline in fertility—an essential condition for activating the demographic dividend, understood as the increase in the proportion of the working-age population relative to the dependent population constitutes the determining and structuring factor of the transition [9, 39, 55, 65, 66]; however, persistently high rates slow the acceleration of economic growth [9, 65]. The realization of this dividend requires simultaneous progress in female education, reduction of infant mortality, strengthening of family planning, and structural transformation of the economy [8, 58, 60]. Societies that combine favorable social environments with robust family planning programs register the most pronounced declines in fertility [58]. Challenges persist, however: slowness or stagnation of the transition, marked intra-regional and national heterogeneity, the uncertain impact of epidemiological crises such as HIV/AIDS, and socioeconomic and cultural barriers—illiteracy or dependence on children in old age—that hinder change [8, 9, 53, 55, 65]. To overcome these risks and harness the potential of the dividend, comprehensive public policies and a nuanced understanding of the region's demographic dynamics are required [8, 55, 58].

7.3 Diversity of Experiences and Final Reflections

After exploring the demographic trajectories of different regions and tracing some common trends through comparative analysis, this final section offers a synthesis of the most relevant structural factors that shape contemporary demographic processes. Although population dynamics manifest profoundly contextually, it is possible to identify cross-cutting patterns, significant regional contrasts, and shared challenges that allow for constructing a more articulated perspective.

The objective of this conclusion is twofold: on the one hand, to systematize the main findings of the chapter through clear, literature-supported comparisons; on the other, to open a reflection on the ethical, political, and social implications of these dynamics, as well as on the need to build adapted, humane, and sustainable public policies. The following sections analyze, in an integrated manner, the key determinants—education, health, economic structures, gender dynamics, and migration—and their differential expression according to the regional context and the normative principles that should guide institutional responses to the global demographic winter.

7.3.1 Factors Related to Demographic Winter

7.3.1.1 Contrasts According to Different Regions

A comparative analysis of regions' main characteristics, challenges, and opportunities is conducted to better understand global demographic trajectories. The following summary table highlights the fundamental contrasts in fertility, aging, and social dynamics, allowing for the visualization of both the diversity of contexts and the shared trends (Table 7.1).

As discussed and detailed in the corresponding table, the world's regions face different demographic challenges. While Europe grapples with accelerated aging and low fertility, Sub-Saharan Africa confronts the challenge of a rapidly growing young population. Latin America and the Arab world are navigating mixed transitions with a double demographic burden, and North America is experiencing moderate aging, mitigated by migratory dynamism. These demographic heterogeneities not only delineate differentiated public policy priorities but also demand contextualized strategic approaches capable of articulating comprehensive responses that capitalize on the opportunities and manage the specific risks of each region.

7.3.1.2 Compared Demographic Dynamics and Regional Contexts

The demographic transition—characterized by the convergence toward low mortality and fertility—is a global phenomenon, although its pace and stage diverge substantially among regions [1, 7]. In Western and Eastern Europe, the combination of

Table 7.1 Key regional demographic trends

Key characteristic	Western/Eastern Europe	Sub-Saharan Africa	LAC/Arab world	North America
Fertility	Very low/drastic decline	High, beginning of slow decline	Heterogeneous, in decline	Low, offset by migration
Aging	Advanced/ accelerated	Young population	Incipient, rapid potential dividend	Significant, growing
Main challenge	Welfare sustainability/ depopulation	Massive investment in youth	Double burden/ heterogeneity	Inequalities/ migrant integration
Key opportunity	Adaptation/ reconstruction	Demographic dividend (with investment)	Demographic dividend/ diversification	Migratory dynamism

Source: Author's elaboration based on the analysis of Sect. 7.2 of the chapter
SSA Sub-Saharan Africa, *LAC* Latin America and the Caribbean, *W.* Western Europe, *E.* Eastern Europe

advanced aging and extremely low fertility predominates [1]. Sub-Saharan Africa, in contrast, faces the challenge of an overwhelmingly young population and the possibility of capitalizing on a demographic dividend that requires heavy investments in human capital and employment [56]. Regions such as Arab Africa, the Middle East, and Latin America are treading heterogeneous paths: they must simultaneously respond to the needs of large youth cohorts and prepare for a rapid increase in the older adult population [6, 7]. North America, whose growth derives mainly from migration, is confronted with an aging process nuanced by profound internal inequalities [43, 44]. The Arab world also exhibits marked national disparities, with countries advancing at different paces and exacerbating socioeconomic gaps [6].

The structural factors that modulate these outcomes can be summarized as follows:

Table 7.2 shows that structural factors—female education, socioeconomic situation, population policies, cultural changes, and migration—do not operate homogeneously: their effect depends on each region's historical, institutional, and cultural conditions. Recognizing these differences is essential for designing effective and context-sensitive demographic policies.

Demographic dynamics reflect a multidimensional interaction that transcends birth and mortality figures. Education, especially for women, emerges as a universal lever for fertility reduction [6, 8, 13, 57, 58], reinforced by the decline in infant mortality and access to family planning and reproductive health [8, 58, 60]. These health priorities differ by context: Europe and North America concentrate on chronic diseases and long-term care [22, 44]; Sub-Saharan Africa prioritizes primary and maternal–child care [56, 58]; and the Arab World faces challenges linked to communicable diseases and nutritional deficits [10].

Economic stability plays an equally decisive role. Recessions reduce birth rates, as in Southern Europe after 2008 [1] and in the transitioning economies of Eastern Europe—while the transformation toward higher-productivity sectors accelerates the decline in fertility, a crucial phenomenon in Sub-Saharan Africa [8, 56]. Political instability also depresses demographic indicators [18]. In parallel, aging strains

Table 7.2 Key structural factors and their regional variation

Structural factor	Main impact on fertility/ development	Main regional differences
Female education	Universally transformative: Reduces fertility, improves health.	Pace of generalization and persistence of illiteracy (Arab world, SSA) vs. high levels (Europe, N. America).
Socioeconomic situation	Crises impact birth rates (Europe); youth unemployment and rent dependency affect (Arab world, SSA).	Stability and development as a prerequisite for transition (SSA) vs. crisis as a disruptor (Eastern Europe).
Population/family policies	From birth control (LAC, SSA) to promotion (Europe) and adaptation (all aging regions).	Need for comprehensive post-crisis policies (Eastern Europe) vs. management of dividends/aging (LAC, Arab world)
Cultural changes (e.g., SDT)	Explain low fertility in the West; echoes in other regions interact with local cultures	Modernity-tradition tension (Arab world) vs. secularization (Europe)
Migration (as a factor)	Offsets low fertility (Europe, N. America); aggravates depopulation (E. Europe); engine of social change	Complex and multidirectional flows (LAC, Africa)

Source: Author's elaboration based on the analysis of Sects. 7.2 and 7.3 of the chapter
SSA Sub-Saharan Africa, *LAC* Latin America and the Caribbean, *E. Europe* Eastern Europe, *N. America* North America, *SDT* Second Demographic Transition

productive, pension, and health systems [1, 4], and their final fiscal impact depends on public policy decisions [4].

Sociocultural and gender factors act as filters that amplify or slow these processes. In the Arab World, improving female status and being open to external ideas are associated with lower fertility [10, 13, 64]. At the same time, in Sub-Saharan Africa, the absence of children carries a strong social stigma [10]. Transversally, gender and class inequalities shape reproductive behavior and access to essential services [7, 56].

Finally, migrations reconfigure global demographic patterns: Western Europe attracts immigration, Eastern Europe experiences notable emigration; North Africa and the Middle East combine outbound and inbound flows; Sub-Saharan Africa combines intraregional mobility with movement to other continents; Latin America maintains its sending profile with relevant internal currents; and in North America, immigration remains the main engine of population growth [43, 44].

7.3.2 Toward Adapted, Humane, and Ethical Policies

The comparative analysis of demographic dynamics highlights the urgency of designing comprehensive public policies that are practical, ethically sustainable, culturally relevant, and socially just. Such policies must be adapted to each context: in Western Europe, the priority is to facilitate the reconciliation between family life and professional activity in the face of growing female educational and labor participation [1, 22]; in Sub-Saharan Africa and parts of Latin America, harnessing the demographic dividend demands strategic investments in education and youth employment [4, 9]; and in aging societies, ensuring intergenerational justice is imperative [1].

Gender equity constitutes a cross-cutting axis, requiring equal access to education, reproductive health, economic opportunities, and decision-making [8, 10]. In Latin America, the decline in fertility has not significantly reduced social inequalities [7]; therefore, policies must be guided by human dignity and justice, avoiding the instrumentalization of people [1]. Likewise, migration management must reconcile economic needs with the complete protection of human rights [43, 52], and any political, financial, or cultural opening must be articulated on solid ethical principles [10, 31].

7.3.3 Conclusion: Weaving the Demographic Future in an Era of Transitions

The "demographic winter" is not uniform but a mosaic of trajectories that requires avoiding generalizations. Underlying it are universal lessons—female education, public health, and family planning, socioeconomic stability—whose effect is reconfigured according to specific contexts: dependency on external rents in the Arab world, the impact of HIV/AIDS in Sub-Saharan Africa, or the postcommunist traumas in Eastern Europe. Far from mere statistics, demographic trends redefine power, governance, and the social contract. When the supranational bodies of Western Europe do not act, welfare systems suffer [21, 27], and in the end, the decisions made by each country weigh more heavily than isolated cultural factors [10]. That is why it is advisable to go beyond short-term electoral logic and plan for the long term: Latin America is aging at great speed [3], and the "demographic dividend" of Sub-Saharan Africa—brief and valuable—requires immediate investment in the development of its people [4, 9, 64].

This challenge is inseparable from ethics. Policies cannot be reduced to adjusting fertility rates or pension balances; they must promote intergenerational justice and gender equity [1, 8], avoiding the instrumentalization of women or migrants without guaranteeing full rights [1, 51]. Managing the "openness" to new ideas and populations [10] requires an ethical framework that places human dignity at the center. The demographic winter is, thus, a call for political and social decisions that guarantee sustainable adaptation and a more equitable future for all generations.

Managing demographic change requires an articulated multilevel governance, where local, national, and international actors assume shared but differentiated responsibilities. Although demographic trajectories are diverse, their effects transcend borders—migration, pressure on health systems, fiscal sustainability, and demand frameworks for global cooperation that balance solidarity and co-responsibility. Adaptation to the "demographic winter" cannot be conceived as an isolated effort by states, but as a collective task that revalues human capital, strengthens democratic institutions, and reactivates social pacts that integrate ethics and sustainability as fundamental axes.

References

1. Gesano G La riproduzione in Italia e nelle sue regioni nel quadro delle dinamiche demografiche in Europa. IRPPS Monogr. 2019. https://doi.org/10.14600/978-88-98822-15-7
2. Zavala de Cosio ME. Démographie, pauvreté et inégalités. Enjeux Dév En Am Lat Dyn Socioéconomiques Polit Publiques. Institut des Amériques. Agence Francaise de Développement; 2011. p. 83–110. Available from: https://shs.hal.science/halshs-00595444.
3. Brenes MC. Orgullosos del asombroso descenso de la fecundidad e inconscientes de la aceleración del envejecimiento (1950–2000). An En Gerontol. 2005;5:83–108.
4. Cotlear D. Envejecimiento de la poblacion ¿Esta preparada America Latina? World Bank; 2011. p. 305.
5. Cosío-Zavala ME. Démographie, pauvreté et inégalités. Agence Française de Développement (AFD); 2011. p. 83–110.
6. La Tabutin D. Démographie du monde arabe et du Moyen-Orient des années 1950 aux années 2000. Population. 2005;60:114.
7. Cosío-Zavala ME. María Eugenia Zavala de Cosío. Demografía, pobreza y desigualdades. Los desafíos del desarrollo en América Latina - Dinámicas socioeconómicas y políticas públicas. Agence Francaise Dév AFD; 2014. Available from: https://shs.hal.science/halshs-00968878v1.
8. Büttner N, Grimm M, Günther I, Harttgen K, Klasen S. The fertility transition in sub-Saharan Africa: the role of structural change. Demography. 2024;61:1585–611.
9. Ursavaş U. Growth acceleration and demographic structure: the case of ECOWAS countries. Fiscaoeconomia. 2022;6:1059–70.
10. Yves Montenay. Démographie politique des pays arabes d'Afrique: la diffusion de la "transition démographique" dans le dernier quart du vingtième siècle: le rôle des facteurs politiques, culturels, religieux et historiques, et leurs interactions avec les évolutions démographiques. These de doctorat, Université Paris-Sorbonne; 1994. Available from: https://theses.fr/1994PA040167.
11. Urrútia G, Bonfill X. Declaración PRISMA: una propuesta para mejorar la publicación de revisiones sistemáticas y metaanálisis. Med Clín. 2010;135:507–11.
12. Abdul Salam A, Elsegaey I, Khraif R, AlMutairi A, Aldosari A. Components and public health impact of population growth in the Arab World. PLoS One. 2015;10:15.
13. Eltigani EE. Fertility transition in Arab countries: a re-evaluation. J Popul Res. 2005;22:163–83.
14. Frini O. Unemployment, divorce, and longevity: the major factors of the fertility upward evolution in Tunisia, during 1998-2018, a dynamic panel data analysis. Hist Fam. 2024;29:373–92.
15. Doignon Y, Blöss-Widmer I, Ambrosetti E, Oliveau S. Population dynamics in the Mediterranean, a demographic convergence? Cham: Springer International Publishing; 2023. https://doi.org/10.1007/978-3-031-37759-4_1.
16. Omar M. The marriage mystery: exploring late marriage in MENA. Al-Raida J. 2001;18:12–9.
17. McKee M, Keulertz M, Habibi N, Mulligan M, Woertz E. Demographic and economic material factors in the MENA region. Work Paper. 2017:43.
18. Mohamed TS. Determinants of fertility rate: a regression analysis case of Tunisia. Univ Marburg; 2016. p. 13.
19. El-Khoury G. Demographic trends and age structural change in Arab countries: selected indicators. Contemp Arab Aff. 2016;9:340–5.
20. Al Zalak Z, Goujon A. Exploring the fertility trend in Egypt. Demogr Res. 2017;37:995–1030.

21. Tosun MS, Yang J. Determinants of fertility and population policies in MENA countries. Econ Res Forum. 2018;42:32.
22. Sardon J-P. Évolution démographique récente des pays développés. Popul Fr Ed. 2004;59:305–60.
23. Van de Kaa DJ. Europe's second demographic transition. Popul Bull. 1987;42:53.
24. Ediev D, Coleman D, Scherbov S. New measures of population reproduction for an era of high migration. Popul Space Place. 2014;20:622–45.
25. Artzrouni MA, Easterlin RA. Birth history, age structure, and post World War II fertility in ten developed countries: an exploratory empirical analysis. Genus. 1982;38:81–99.
26. Thalberg S. Demographic patterns in Europe. A review of Austria, Germany, The Netherlands, Estonia, Latvia, and Lithuania. Arbetsrapport Institutet För Fram.; 2003. p. 65.
27. Van De Kaa DJ. Europe's second demographic transition. Popul Bull. 1987;42:1–59.
28. Hülskamp NE. Ursachen niedriger Fertilität in hoch entwickelten Staaten. Soziologische, ökonomische und politische Einflussfaktoren. Köln: Kölner Univ.-Verl; 2006.
29. Pierrakos G, Balourdos D, Soulis S, Sarris M, Pateras J, Skolarikos P, et al. Comparative analysis and evaluation of the effectiveness of demographic policies in EU countries (2009–2010). World Health Popul. 2014;15:64.
30. Tønnessen M, Kravdal Ø, Syse A, Skjerpen T, Cappelen Å, Carling J. Demographic consequences of migration. Univ Oslo; 2017. p. 86.
31. Jemna D-V, Cigu E. Analysis of fertility in ten central and eastern European countries after 1989. Transylv Rev Adm Sci. 2014;10:49–77.
32. Hoff A. Population ageing in central and eastern Europe as an outcome of the socio-economic transition to capitalism. Soc Inq Well-Being. 2008;7:14–25.
33. Kotowska IE, Jóźwiak J, Smith PV, Matysiak A, Muszyńska M. Population of central and Eastern Europe. Challenges and opportunities. Institute of Statistics and Demography at the Warsaw School of Economics; 2003. p. 734.
34. Billingsley S. The post-communist fertility puzzle. Popul Res Policy Rev. 2010;29:193–231.
35. Jemna D-V, David M. Fertility decline and socio-economic uncertainty in central and eastern european countries after 1990. Appl Econ. 2022;54:3296–308.
36. Zastavetska L, Zastavetskyi T. Socio-geographical dimension of demographic processes in Ukraine in the period of independence (from 1991 to 2022). Sci Issues Ternopil Volodymyr Hnatiuk Natl Pedagog Univ Ser Geogr. 2022;52:66–74.
37. Collver A, Arevalo JV. Current trends and differentials in fertility as revealed by official data. Milbank Mem Fund Q. 1968;46:39.
38. Obradoviæ J, Èudina M. Child-bearing attitudes in Croatia: a country passing through the modernization process. J Comp Fam Stud. 2011;42:77–93.
39. Affairs D of E and S. World population ageing 2023: Challenges and opportunities of population ageing in the least developed countries. 1st ed. Bloomfield: United Nations Research Institute for Social Development; 2024.
40. Доманчук Т, Грицюк М, Чорненька Ж, Марараш Г, Собко Д. Current trends in the dynamics of the demographic situation in Ukraine and in the Chernivtsi region, particularly for 2001–2021. Перспективи Та Інновації Науки; 2023.
41. Gendell M. Fertility and development in Brazil. Demography. 1967;4:143–57.
42. Castañeda T. Determinantes del Cambio Poblacional en Colombia. Rev Desarro Soc. 1980:319–34.
43. Barbieri M, Ouellette N. The demography of Canada and the United States from the 1980s to the 2000s: a summary of changes and a statistical assessment. Popul Engl Ed. 2012;67:177.
44. Butz W, McCarthy KF, Morrison PA, Vaiana ME. Demographic challenges in America's future. Rand Corporation; 1982. Available from: https://www.rand.org/pubs/reports/R2911.html.
45. Munnell A, Chen A, Sanzenbacher G. Is the drop in fertility due to the great recession or a permanent change? SSRN Electron J. 2019. https://doi.org/10.2139/ssrn.3362478
46. Vining Jr DR. Below-replacement fertility in five regions of the world. Mank Q. 1989;29:203–20.

47. Bauch CT. Wealth as a source of density dependence in human population growth. Oikos. 2008;117:1824–32.
48. Leeson GW. Global demographic change and the case of low fertility. Popul Horiz. 2018;15:1–6.
49. Anderson TM. Three essays on the social, economic, and demographic causes and consequences of low fertility. Pennsylvania: Faculties of the University of Pennsylvania; 2015. Available from: https://repository.upenn.edu/entities/publication/2cbea3bc-4f87-4cea-88fb-a15826fd526e.
50. Espenshade TJ. Population dynamics with immigration and low fertility. Popul Dev Rev. 1986;12:248.
51. The Northern America Fertility Divide Hoover Inst. [cited 2025 May 7]. Available from: https://www.hoover.org/research/northern-america-fertility-divide.
52. Espenshade TJ. Zero population growth and the economies of developed nations. Popul Dev Rev. 1978;4:645–80.
53. Linder A. Analysis of the determinants of the fertility decline in southern Africa [Master Programme in Economic Demography]. Lund University; 2015. Available from: http://lup.lub.lu.se/student-papers/record/7521804.
54. Potts D, Marks S. Fertility in southern Africa: the quiet revolution. J South Afr Stud. 2001;27:189–205.
55. Teller CH, Hailemariam A, Gebreselassie T, Seifu Y. The uniqueness of the Ethiopian demographic transition within sub-Saharan Africa: multiple responses to population pressure and preconditions for rural fertility decline and capturing the demographic dividend. Afr Popul Stud. 2011;25:362–80.
56. Shapiro D, Gebreselassie T. Fertility transition in sub-Saharan Africa: falling and stalling. Afr Popul Stud. 2013;23:10.11564/23-1-310.
57. Shapiro D, Hinde A. Laggards in the global fertility transition. Vienna Yearb Popul Res. 2020;18:123–40.
58. Jain AK, Ross JA. Fertility differences among developing countries: are they still related to family planning program efforts and social settings? Int Perspect Sex Reprod Health. 2012;38:015–22.
59. Chisadza C, Bittencourt M. The Fertility Transition: Panel Evidence from sub-Saharan Africa. Econ Res South Afr ERSA Res Programme Funded Natl Treas South Afr. 2016.; Available from: https://api.semanticscholar.org/CorpusID:157934299.
60. Shapiro D. Women's education and fertility transition in sub-Saharan Africa. Vienna Yearb Popul Res. 2013;10:9–30.
61. Shapiro D, Tenikue M. Women's education, infant and child mortality, and fertility decline in urban and rural sub-Saharan Africa. Demogr Res. 2017;37:669–708.
62. Scribner S. Policies affecting fertility and contraceptive use: an assessment of twelve sub-Saharan countries. The World Bank; 1995. Available from: http://documents.worldbank.org/curated/en/922431468767674901.
63. Mahy M, Gupta N. Trends and differentials in adolescent reproductive behavior in sub-Saharan Africa. Calverton: ORC Macro; 2002. Report No.: 3. Available from: http://dhsprogram.com/pubs/pdf/AS3/AS3.pdf.
64. Verkroost FCJ, Monden CWS. Childlessness and development in sub-Saharan Africa: is there evidence for a U-shaped pattern? Eur J Popul. 2022;38:319–52.
65. Amarachi U-A. Essays on fertility and the demographic dividend in sub-saharan Africa: exploring the puzzles and possibilities. Universtiy Pennsylvania; 2018. Available from: https://repository.upenn.edu/edissertations/2934.
66. Li Q, Tsui AO, Liu L, Ahmed S. Mortality, fertility, and economic development: an analysis of 201 countries from 1960 to 2015. Gates Open Res. 2018;2:14.

Chapter 8
Global Governance and Public Health

Alberto García Gómez and Martina Todaro

8.1 Introduction

Public health has established itself as one of the pillars of stability and human development on a global scale. However, contemporary challenges—pandemics, health inequalities, ecological crises, mass migrations, and geopolitical tensions—have highlighted the fragilities of health systems and the urgent need for ethical, responsible, and pluralistic governance. In this context, a profound reflection is needed on the moral foundations that should guide public health decision-making and on the necessity of integrating diverse cultural, philosophical, and religious perspectives in the formulation of equitable and sustainable public policies.

The UNESCO Universal Declaration on Bioethics and Human Rights (2005) provides a normative framework guided by universal ethical principles, such as human dignity, informed consent, equity, social justice, and solidarity. However, effectively implementing these principles in the plural realities of the twenty-first century requires an intercultural and interreligious dialogue that favors the local and contextualized appropriation of universal values. The philosophical and religious traditions of Buddhism, Confucianism, Christianity, Hinduism, Islam, Judaism, and Taoism, rather than being obstacles, can offer profound ethical and spiritual resources to support more humane, compassionate, and committed governance for the common good.

This chapter aims to reflect on the ethics of governance in public health from an intercultural and interreligious perspective, articulating the principles of the UNESCO Declaration with the values promoted by the Social Teaching of the Church and by various religious worldviews. To this end, the contributions of each

A. García Gómez (✉) · M. Todaro
UNESCO Chair in Bioethics and Human Rights, Ateneo Pontificio Regina Apostolorum and Università Europea di Roma, Rome, Italy
e-mail: alberto.garcia@upra.org; mtodaro@unescobiochair.org

D. Cerdio-Domínguez, M. E. de los Ríos-Uriarte (eds.), *Current Bioethics*, New Paradigms in Healthcare, https://doi.org/10.1007/978-3-032-25294-4_8

tradition to bioethical thinking will be addressed, the conceptual convergences between them will be analyzed, and practical guidelines for ethical, supportive, and subsidiary governance will be proposed. In addition, contemporary contributions from virtue ethics and global bioethics will be incorporated to enrich the dialogue with the challenges of the present.

The ultimate purpose of this reflection is to open paths toward a governance ethic that is not only normative but also spiritually inspiring and culturally resonant, capable of generating trust, co-responsibility, and a sense of ownership in public health policies.

8.2 UNESCO's Universal Declaration on Bioethics and Human Rights as an Ethical Framework for Public Health Governance

The UNESCO Universal Declaration on Bioethics and Human Rights (2005) is one of the most ambitious and globally representative attempts to establish a common ethical framework for life sciences, health, environment, and human rights. Unlike other normative instruments focused on biomedical ethics, this Declaration articulates bioethics with the fundamental principles of international human rights law, underlining the interdependence between human dignity, autonomy, solidarity, social justice, and shared responsibility in the field of public health.

In its Preamble, the Declaration states that "health does not depend solely on scientific and technological research developments but also on psychosocial and cultural factors" [1].

This integrative perspective enables overcoming reductionist approaches, opening up space for a holistic understanding of health as a public good and human right, rather than merely as a technical result or market commodity.

8.2.1 Key Principles Applicable to Governance

The Declaration contains 15 ethical principles that guide decision-making in bioethics. Of these, several are particularly relevant to public health governance:

- Human dignity and human rights (Art. 3.1): "Human dignity, human rights and fundamental freedoms are to be fully respected".
- The person must be recognized as an end in itself, never as a means, which implies policies centered on integral respect for the human being.
- Justice and equity (Art. 10): "The fundamental equality of all human beings in dignity and rights is to be respected so that they are treated justly and equitably."

- Equitable distribution of the benefits of science and nondiscriminatory access to health services are required.
- Solidarity and cooperation (Art. 13): "Solidarity among human beings and international cooperation towards that end are to be encouraged."
- Collective action must guide global governance in the face of shared challenges, such as pandemics, climate change, or forced migration.
- Individual and social responsibility (Art. 14): "Taking into account that the enjoyment of the highest attainable standard of health is one of the fundamental rights of every human being without distinction of race, religion, political belief, economic or social condition, progress in science and technology should advance."

Recognizes the duty of States and public actors to generate environments that promote health and well-being.

This body of norms not only defines what is desirable but also establishes concrete ethical obligations that should guide the health governance architecture, taking into account both local contexts and universal principles.

8.2.2 The Participatory and Cultural Dimension of Ethics

One of the most innovative contributions of the Declaration is its openness to the plurality of cultural and religious traditions. Article 12, dedicated to respect for cultural diversity, states that "the importance of cultural diversity and pluralism should be given due regard." This principle recognizes that the legitimacy of health policies does not depend exclusively on scientific evidence but also on their moral resonance in the contexts where they are applied.

As Darlei Dall'Agnol points out, this ethical openness does not imply a relativism without limits, but a "dialogical universalism" that seeks common principles without ignoring differences [2]. Hence, the relevance of an approach that articulates the Declaration with the ethical doctrines of various religions will be developed in the following chapters.

8.2.3 UNESCO and Global Governance: Guiding Principles

The notion of global governance in public health refers to a network of actors—state, multilateral, scientific, and community—that must cooperate in the definition, implementation, and evaluation of equitable and sustainable health policies. UNESCO proposes a normative vision in which ethical governance is not reduced to the technical design of strategies but promotes:

- Informed participation of all stakeholders.
- Transparency in decision-making processes.

- Evaluation of the ethical impact of health policies.
- Intercultural and interreligious dialogue as a means of ethical legitimization.

These guidelines coincide with those of authors such as Ruth Macklin, who argues that global health ethics must integrate universalist principles with local sensitivity, recognizing that cultures interpret health in different ways [3].

8.3 Ethical Convergence between the UNESCO Declaration and the Social Teaching of the Church

8.3.1 Introduction to the Social Teaching of the Church (SDC)

The Social Teaching of the Catholic Church (SDC) constitutes a systematic body of moral and social reflection that, from the ecclesial magisterium, seeks to illuminate contemporary social realities with the light of the Gospel. Born with the encyclical *Rerum Novarum* (1891) of Leo XIII [4] and progressively developed by subsequent pontiffs, the SDC articulates principles such as human dignity, the common good, solidarity, and subsidiarity as axes of analysis and ethical orientation.

These principles offer a profoundly humanistic and relational vision of the human being, in which the person is not an isolated individual, but an essentially social being whose dignity is rooted in his or her openness to others and God. This conception is convergent with the relational and communitarian approach that underlies the UNESCO Universal Declaration on Bioethics and Human Rights.

8.3.2 Basic Coincidences: Dignity, Common Good, and Justice

8.3.2.1 Human Dignity as a Foundation

Both the UNESCO Declaration and the SDC recognize the intrinsic and inviolable dignity of every human being as a nonnegotiable principle. The encyclical *Caritas in Veritate* (2009) of Benedict XVI [5] affirms that the dignity of the human person and respect for human rights cannot depend on consensual or pragmatic decisions but on the truth of the human being.

UNESCO, for its part, in Article 3 of its Declaration, stresses that "human dignity, human rights and fundamental freedoms must be fully respected."

8.3.2.2 The Common Good as a Horizon for Public Action

The SDC defines the common good as "the sum total of social conditions which allow people, either as groups or as individuals, to reach their fulfillment more fully and more easily" [6]. This notion is profoundly ethical and political, and is fully in line with article 14 of the UNESCO Declaration, which underlines the social responsibility to ensure conditions for the well-being of all.

Pope Francis, in *Fratelli tutti*, takes up this idea forcefully: "As a community, we have an obligation to ensure that every person lives with dignity and has sufficient opportunities for his or her integral development" [7].

8.3.2.3 Distributive Justice and Equity in Health

Both frameworks agree on the obligation to guarantee fair and equitable access to health resources. The SDC insists on the "universal destination of goods" and denounces structural inequalities as ethical and spiritual violations. In the health context, this implies that life and health cannot be commodities subject to the market but are universal rights. The Compendium of the SDC states:

> 177. Christian tradition has never recognized the right to private property as absolute and untouchable: "On the contrary, it has always understood this right within the broader context of the right common to all to use the goods of the whole of creation: the right to private property is subordinated to the right to common use, to the fact that goods are meant for everyone" [8]. The principle of the universal destination of goods is an affirmation both of God's full and perennial lordship over every reality and of the requirement that the goods of creation remain ever destined to the development of the whole person and of all humanity. This principle is not opposed to the right to private property but indicates the need to regulate it. Private property, in fact, regardless of the concrete forms of the regulations and juridical norms relative to it, is in its essence only an instrument for respecting the principle of the universal destination of goods; in the final analysis, therefore, it is not an end but a means [9].

8.3.3 Subsidiarity and Inclusive Governance

The principle of subsidiarity, which establishes that decisions should be made as close as possible to the people affected, is closely related to Article 6 of the UNESCO Declaration, which promotes free and informed consent as an expression of individual and community autonomy.

Applied to public health governance, this principle requires:

- Strengthening of local capacities.
- Respect for community authorities.
- The articulation between the local and the global in a complementary manner.

The theologian Gustavo Gutiérrez, one of the pioneers of Latin American liberation theology emphasized, in a period challenged by his thinking, that there can be no option for the poor without structural inclusion and active participation in the decisions that affect their health and lives: "Poverty is the threat and the reality of an unjust and early death" and the option "involves standing in solidarity with the poor and a stance against inhumane poverty" [10].

8.3.4 *Solidarity as a Cross-Cutting Theme*

Solidarity, understood as an active commitment to others, especially the most vulnerable, occupies a central place in both the SDC and the UNESCO Declaration. The latter, in its Article 13, states: "Solidarity among human beings and international cooperation towards that end are to be encouraged."

John Paul II, in Sollicitudo Rei Socialis, 1987, no. 38, defines solidarity as: "a firm and persevering determination to commit oneself to the common good; that is to say to the good of all and of each individual, because we are all really responsible for all" [11].

8.4 Religious and Cultural Traditions in Dialogue with Public Health Governance

Public health and its global ethical governance require an intercultural and interreligious openness that recognizes the knowledge and convictions of diverse traditions. The world's religions—in their philosophical, spiritual, and ethical dimensions—offer deeply rooted principles of dignity, compassion, justice, and care for the other. This section examines the main contributions of the traditions of Buddhism, Confucianism, Christianity, Hinduism, Islam, Judaism, and Taoism, identifying their convergences with the principles of the UNESCO Universal Declaration on Bioethics and Human Rights and the Social Teaching of the Church.

8.4.1 *Buddhism: Compassion, Interdependence, and Suffering*

Buddhism, with its central principle of karuṇā (compassion), places the alleviation of suffering as a moral priority. According to His Holiness the Dalai Lama, leading a compassionate life means that "one naturally becomes deeply engaged and committed to be involved with and care about other people" [12]. The idea of duḥkha (suffering) as an inherent reality of existence justifies ethical action in public health from care, prevention, and compassion.

Furthermore, the notion of interdependence (pratītyasamutpāda) resonates with the principle of solidarity and the common good. In the context of the COVID-19 pandemic, Buddhist thinkers such as Damien Keown have highlighted how Buddhist compassion can inspire health policies focused on equity and the prevention of unnecessary suffering [13].

8.4.2 *Confucianism: Social Harmony and Relational Responsibility*

Confucianism emphasizes human relationships and the duty to others, based on the principles of ren (benevolence) and li (social rituals that govern life). According to Ruiping Fan: "Confucian moral thought regards persons in general and patients in particular as located within families because Confucian thought appreciates the deeply family-centered character of social reality" [14].

In Confucian thought, the virtuous ruler (junzi) promotes just health policies, inspired by an ethic of hierarchical but just care. In this sense, it aligns with subsidiarity, which holds that decisions should be made with respect for local and community ties. Likewise, the search for harmony (he) is linked to the vision of the common good and solidarity.

8.4.3 *Christianity: Inalienable Dignity and Option for the Poor*

The Christian tradition, as presented in the Social Teaching of the Church and moral theology, emphasizes the dignity of the human being as the image of God (*imago Dei*).

The Gospel message contains a clear imperative: "For I was hungry and you gave me something to eat, I was thirsty and you gave me something to drink, I was a stranger and you invited me in, I needed clothes and you clothed me, I was sick and you looked after me, I was in prison and you came to visit me." (Mt 25:35-36), which underlies an ethic of care and active solidarity.

In addition, the preferential option for the poor, cultivated by liberation theology and reaffirmed by Pope Francis, calls for inclusive governance, focusing on the most vulnerable. This tradition converges with the principle of global health equity and the social responsibility of States.

8.4.4 *Hinduism: Dharma, Integral Health, and Common Destiny*

In Hinduism, the concept of dharma (moral order and ethical duty) includes the individual's responsibility toward the community and nature. In healthcare, dharma involves protecting life, promoting balance between body, mind, and spirit (*trividha tapa*), and supporting health as a common good.

Ayurveda, the traditional medical system, promotes a holistic, community-based, preventive approach to health, in keeping with the principles of sustainability and respect for life. Cosmic interconnectedness, expressed in the phrase "Vasudhaiva Kutumbakam" ("The world is one family"), underpins an ethic of global governance that resonates with the principles of interdependence and shared responsibility.

8.4.5 Islam: Justice, Mercy, and the Common Good (Maṣlaḥa)

Islam promotes a vision of health grounded in justice (ʿadl), mercy (raḥma), and the common good (maṣlaḥa). The Qur'an states: "Whoever saves one life, it is as if they have saved all mankind" (Qur'an 5:32). This ethical imperative underpins the collective duty to ensure equitable access to healthcare for all.

The concept of maqāṣid al-sharīʿa (higher ends of Islamic law) includes the protection of life and health. Ethicists such as Abdel Rahman Al-Awadi have insisted that health systems should be geared to the general welfare and not to corporate [5] interests. Likewise, the principle of taʿāwun (cooperation) embodies active solidarity with direct implications for the governance of pandemics and vaccine policies.

8.4.6 Judaism: Tikkun Olam, Justice, and Care for the Vulnerable

Judaism teaches the notion of tikkun olam (repairing the world) as an ethical responsibility to improve and heal society. The Talmud states: "Whoever saves a single life, it is as if they have saved an entire world" (Sanhedrin 37a), a maxim that has inspired numerous Jewish initiatives in public health and social justice.

Moreover, the principle of tzedakah (social justice) states that helping those in need is an obligation, not an option. For thinkers such as Lord Jonathan Sacks, Jewish ethics demands structures of social care and collective responsibility, particularly in addressing illness and poverty, grounded in the principle of tzedakah as a moral obligation to repair society [15]. The community is seen as the guarantor of the health and dignity of its members, in line with the principles of solidarity and equity.

8.4.7 Taoism: Harmony with Nature and Health as Balance

Taoism, centered on the concept of the Dao (Way), promotes harmony between humans and nature as a fundamental principle of health and well-being. The classic text Dao De Jing emphasizes flowing with the natural order and preserving balance, as reflected in passages such as: "The world is a sacred vessel that cannot be changed. He who changes it will destroy it. He who seizes it will lose it." (Dao De Jing, Chap. 29). The ideal of wu wei (non-action or effortless action) invites a mode of living and governance that avoids forced interference, which can be interpreted as a call for noninvasive yet profoundly preventive approaches in public health policies.

Taoism advocates nurturing both the body and the environment, reflecting a deep respect for the natural rhythms of life. Its holistic perspective highlights the importance of harmony between humans and the natural world, resonating strongly with modern principles of environmental stewardship and sustainable health. In this respect, Taoist principles anticipate the "One Health" framework advocated by the World Health Organization (WHO), which recognizes the interconnectedness of human, animal, and ecosystem health.

8.4.8 *Cross-Cultural Convergences and Synthesis*

The religious traditions examined show several notable convergences:

- Human dignity as a nonnegotiable value (present in Christianity, Hinduism, Islam, and Judaism).
- Compassion as a central principle of ethical action (Buddhism, Confucianism, Christianity, and Islam).
- Distributive justice and commitment to the poor (Christianity, Hinduism, Islam, and Judaism).
- The relational and interdependent vision of the human being (Confucianism, Christianity, Buddhism, and Taoism).
- Emphasis on the common good and social harmony (Confucianism, Christianity, Islam, and Taoism).
- Respect for nature and integral health (Christianity, Taoism, Hinduism, Buddhism).

These convergences offer fertile ground for a global ethic of public health governance that is sensitive to cultural and spiritual differences but also grounded in universal values.

8.5 Ethics Applied to Global Health Governance: Proposals from Interculturality and Bioethics

Cultural and religious diversity should not be perceived as an obstacle to global health governance, but as a source of ethical and conceptual richness. This chapter develops concrete proposals that integrate the above reflections, showing how an intercultural ethic can be applied to contemporary global public health challenges. It articulates a vision of governance based on shared principles, with special emphasis on justice, solidarity, participation, and respect for diversity.

8.5.1 Universal Principles and Plural Foundations

Since the Universal Declaration on Bioethics and Human Rights [1], it is recognized that there are common principles—such as human dignity, consent, justice, equity, and solidarity—that can find diverse expressions in different cultures. The proposal is not to impose a homogeneous model, but to weave a plural bioethical convergence, in which each tradition contributes its understanding of good and care.

In this framework, the principles of interdependence, common good, subsidiarity, and democratic participation—central to the Social Teaching of the Church and various religious traditions—provide a solid basis for a culturally sensitive and socially inclusive ethic of health governance.

8.5.2 Participatory and Intercultural Governance

The COVID-19 pandemic demonstrated that vertical, technocratic, and Western-centered decision-making models were not always effective or ethically valid. The implementation of health policies that did not consider local culture, religions, or traditional forms of care generated resistance, mistrust, and inequality.

Therefore, participatory health governance is proposed, based on:

- Intercultural consultation: incorporate community leaders, religious leaders, women, indigenous peoples, and minority groups in the design and evaluation of health policies.
- Cultural adaptation of interventions: respecting local forms of care, festivities, values, and languages.
- Community bioethics education fosters dialogue between local scientific and ethical knowledge to empower citizens. This approach aligns with the World Health Organization's call for "community engagement" strategies, which emphasize involving communities in health decision-making to build trust and promote equity. It also reflects the recommendations of the Global Preparedness Monitoring Board (2021), which highlight the importance of public trust and equitable access in global health preparedness and response.

8.5.3 Global Justice and Shared Responsibility

The unequal distribution of vaccines, treatments, and health resources has highlighted the need for an ethic of global justice. COVID-19 vaccines, for example, were initially concentrated in the countries of the Global North, while large sectors of the Global South were marginalized.

Ethical governance should include:

- Binding mechanisms for equitable distribution of health resources.
- Transfer of technology and know-how to developing countries.
- Condonation of health debts in humanitarian emergency contexts.
- Multilateral commitments of sustained funding to strengthen health systems in impoverished countries.

These actions are based on the principle of global solidarity, which is widely supported by Buddhist, Christian, Muslim, and secular humanist thought.

8.5.4 Caring for the Most Vulnerable: An Essential Ethical Axis

Intercultural ethics emphasizes the moral obligation to care for the most vulnerable: the poor, the elderly, migrants, persons with disabilities, and indigenous peoples. This priority is based on both religious arguments (preferential option for the poor, compassion, charity, and justice) and bioethical principles (equity and non-maleficence).

In terms of public policy, this implies:

- Guaranteeing universal access to health services, especially in rural and marginalized areas.
- Protecting the rights of migrants and refugees, without discrimination.
- Ensuring culturally appropriate access to health information and preventive services.
- Investing in mental health and palliative care, often invisible.

8.5.5 Planetary Health: A Shared Moral Responsibility

Ethics applied to governance cannot ignore the ecological dimension. Human health is interconnected with the health of ecosystems, as demonstrated by the spread of zoonoses linked to deforestation and animal trafficking. For this reason, a vision of planetary health is being promoted, inspired by the "One Health" approach, but reinforced by religious convictions about the care of creation. Pope Francis, in *Laudato si'*, calls for an "integral ecology" that overcomes the technocratic paradigm.

Indigenous and Eastern traditions also emphasize the interdependence of all beings.

Promoting policies that protect biodiversity, reduce polluting emissions, and promote sustainable lifestyles is, in this context, a global ethical imperative.

8.5.6 Final Proposal: An Ethics of Global Care

This chapter concludes with the articulation of an integrated ethical proposal for global health governance:

- Focused on human dignity and the common good.
- Sensitive to cultural and religious differences.
- Committed to justice, equity, and participation.
- Responsible for the future of human life and the planet.

The ethics of global care is not a closed model, but a shared horizon that must be permanently built-in dialogue, humility, and listening.

8.6 Conclusion: Toward a Global Bioethics with a Vocation for Peace and Health Justice

In a world marked by the interdependence of peoples, cross-border risks, and shared vulnerabilities, the horizon of truly ethical public health governance is not only desirable but urgent. Recent health crises—from pandemics to environmental catastrophes—have highlighted both the limits of fragmented approaches and the transformative power of a response based on mutual care, international solidarity, and respect for the dignity of all.

This chapter has sought to offer a comprehensive, intercultural, and interreligious vision of the ethics of global health governance, in the light of the UNESCO Universal Declaration on Bioethics and Human Rights, in dialogue with the great philosophical and spiritual traditions of the world: Buddhism, Confucianism, Hinduism, Islam, Judaism, and Taoism. Despite their doctrinal differences, they all converge in some fundamental values: compassion, justice, the search for truth, respect for life, the centrality of the person, harmony with nature, and concern for the common good.

Global bioethics, in this sense, presents itself not as a uniform ideology, but as a living platform for the encounter of knowledge, cultures, and experiences. Its vocation is humanistic, open to reason and dialogue, and deeply committed to a vision of health that is not only technical but also moral, ecological, and spiritual. Adela Cortina emphasizes that the ethics of care is not a sentimental or soft option but a rational demand for justice grounded in the universal human condition of vulnerability. This ethics calls for social responsibility and structures that protect and empower the most vulnerable members of society [16, 17].

Bioethics of the common good, inspired by an open reason, as proposed by Benedict XVI and continued by Pope Francis, has the potential to found a new political rationality for global health. It is a reason that does not confine itself to technocratic or utilitarian paradigms, but is open to the great questions of meaning, to sapiential traditions, to listening to the other. Martha Nussbaum emphasizes that a just society depends not only on laws but also on the capacity for empathetic imagination—the ability to understand and share the feelings of others, which is essential for genuine social justice [18].

This path also requires an integral ecology, in which the health of human beings cannot be separated from the health of the Earth. Environmental bioethics, far from being a marginal specialization, must be at the center of global health policy. Resource exploitation, environmental degradation, biodiversity loss, and climate change directly affect the living conditions and health of the poorest and most vulnerable. As Laudato si' reminds us, everything is connected. Social justice and ecological justice are not parallel paths, but one and the same.

Finally, this proposal is part of a vision of governance as a service to peace. Governance that listens, dialogues, cares for, and promotes the integral development of peoples becomes a privileged instrument of reconciliation, stability, and hope. In the words of Hans Küng: "There will be no peace among nations without peace among religions, and there will be no peace among religions without dialogue among them" [19]. And we could add: there will be no health without justice, and there will be no justice without an ethic that unites the local with the universal, the technical with the human, the scientific with the spiritual.

Therefore, we trust that this ethics of governance, rooted in a global, intercultural bioethics and open to integral ecology, may be one of the most fruitful ways to advance toward an inclusive, equitable, sustainable, and peace-promoting public health. Fostering the art of convergence and cooperation in global bioethics of encounter—affirmative and transformative—inspired by expanded reason and love for one's neighbor makes as to offer responses to the ethical challenges of a world marked by cultural, epistemic, and social polarization and seeks to open pathways for dialogue, cooperation, and convergence in contexts of diversity and conflict.

Building bridges among different forms of knowledge, cultures, and convictions and opening borders in our relationship with others (as recently stated Leo XIV) [20] to promote a global bioethics approach that is open and welcoming are nowadays a moral imperative: is needed, is useful, and also beautiful to move toward peace. Global polarization—political, cultural, epistemic, and moral—is eroding our collective ability to respond ethically to humanity's greatest challenges. This fragmentation weakens shared moral languages, undermines the construction of solid ethical consensus, creates gaps in the protection of the human person and peoples, and reduces public deliberation to defensive or utilitarian positions.

In the face of this crisis of dialogue and meaning, we seek to recover an open, affirmative, and deeply human bioethics—one that upholds the value of life, the dignity of every person, and our interdependence with our common home. In this way, we aim to address contemporary dilemmas through a pedagogy of encounter, offering tools to rebuild relationships, restore ethical trust, and open processes of interdisciplinary and intercultural convergence.

References

1. UNESCO. Universal Declaration on Bioethics and Human Rights. Paris: UNESCO; 2005.
2. Dall'Agnol D. Ética global e interculturalidade: desafios para a bioética. Veritas. 2012;57(3):109–26.
3. Macklin R. Ethics in Global Health: principles for global governance. Cambridge: Cambridge University Press; 2012.
4. Leo XIII. Rerum Novarum [Encyclical letter]. Vatican.va. Available from: https://www.vatican.va/content/leo-xiii/en/encyclicals/documents/hf_l-xiii_enc_15051891_rerum-novarum.html
5. Benedict XVI. Caritas in Veritate. Vatican City: Libreria Editrice Vaticana; 2009.
6. Second Vatican Ecumenical Council. Pastoral Constitution. Gaudium et Spes Acta Apostolicae Sedis. 1966;58:1046.
7. Francisco. Fratelli tutti. Vatican City: Libreria Editrice Vaticana; 2020.
8. John Paul II. Encyclical Letter Laborem Exercens: Libreria Editrice Vaticana; 1981;14: AAS 73:613.
9. Pontifical Council for Justice and Peace. Compendium of the social doctrine of the church. Vatican City: Libreria Editrice Vaticana; 2004.
10. Gutiérrez G. Theology from the experience of the poor: a talk delivered by Father Gutiérrez at the 1992 convention of the Catholic Theological Society of America in Pittsburgh. Cathol Theol Soc America: Pennsylvania; 1992.
11. John Paul II. Sollicitudo Rei Socialis: Libreria Editrice Vaticana; vol. 38; 1987.
12. Snyderman R, Gyatso T. Compassion and health care: a discussion with the Dalai lama. Acad Med. 2019;94(8):1068–70. https://doi.org/10.1097/ACM.0000000000002709.
13. Keown D. Buddhist ethics: a very short introduction. Oxford University Press; 2005.
14. Fan R. Confucian familism and its bioethical implications. In: The family, medical decision-making, and biotechnology. Philosophy and medicine, vol. 91. Dordrecht: Springer; 2007. p. 25–40. https://doi.org/10.1007/1-4020-5220-0_2.
15. Kellner M. To heal a fractured world: the ethics of responsibility (review). Shofar Interdiscip J Jew Stud. 2009;26:234–5. https://doi.org/10.1353/sho.0.0010.
16. Cortina A. The experience of vulnerability in ethics in care services for people with severe mental disability. Fundació Grifols; 2015.
17. Cortina A. Cosmopolitan ethics. Rev Latinoam Bioét. 2022;22(1):50–60.
18. Nussbaum MC. Upheavals of thought: the intelligence of emotions. Cambridge University Press; 2001.
19. Küng H. Address at the opening of the exhibit on the world's religions. Santa Clara University; 2005.
20. Leo XIV. Homily Sunday Pentecost. Vatican.va; 2025. Available from: https://www.vatican.va/content/leo-xiv/en/events/event.dir.html/content/vaticanevents/en/2025/6/8/pentecoste.html.

Chapter 9
Beyond Borders: Global Bioethics and the Humanization of Migration Policy

María Victoria Fernández Molina and José Rafael Grijalva Eternod

9.1 Introduction

Migration flows have been a constant feature throughout human history, driven by environmental, economic, social, and political factors. From the dispersal of Homo sapiens out of Africa approximately 100,000 years ago to the ongoing evolution of the state into the contemporary era, migration has shaped the development of civilizations and the distribution of populations across the globe. Over time, migratory patterns have evolved from nomadic movements in search of resources to organized flows influenced by agriculture, trade, warfare, and the expansion of major empires [1].

During antiquity, empires such as the Egyptian, Persian, Greek, and Roman facilitated both voluntary migrations and forced displacements arising from military conquests, commerce, and slavery [2]. With the European expansion, which began in the fifteenth century, migration took on a transcontinental character, marked by the African slave trade and the forced displacement of Indigenous populations in the Americas. Paradoxically, this period also saw theoretical debates that legitimized European migration waves: the *ius migrandi*—first proposed by Francisco de Vitoria

M. V. F. Molina (✉) · J. R. G. Eternod
Inter-American Academy of Human Rights (AIDH), Universidad Autónoma de Coahuila (UAdeC), Mexico City, Mexico
e-mail: victoria.fernandez@uadec.edu.mx; jose.grijalva@uadec.edu.mx

D. Cerdio-Domínguez, M. E. de los Ríos-Uriarte (eds.), *Current Bioethics*, New Paradigms in Healthcare, https://doi.org/10.1007/978-3-032-25294-4_9

(1539)[1] as a natural right of movement and commerce, later developed by Hugo Grotius (1625)[2]—was instrumentalized to justify European settlement in the Americas. While colonists invoked this principle to migrate freely, these same theorists denied its application to Indigenous and enslaved peoples, institutionalizing a racialized double standard [3].

This conceptual framework laid the groundwork for the consolidation of the modern state in the seventeenth and eighteenth centuries, when human mobility transitioned from a spontaneous phenomenon to a tightly regulated process. European governments actively promoted colonial migration through subsidies and indentured labor contracts while criminalizing the movement of non-European populations, thereby establishing the legal foundations of a global migration control system rooted in ethnic and economic hierarchies [4]. This framework, though dressed in the language of universal rights, ultimately served to enforce colonial domination under the guise of state sovereignty [5].

The construction of the Westphalian nation-state introduced a new legal and political paradigm based on territorial sovereignty, establishing clearly defined borders and systems for controlling migration. A key function of the modern sovereign state is population management. As Torpey (1998) has noted, modern nation-states not only have a monopoly over the legitimate means of violence but also have the legitimate means of movement. Their territorial logic means that they have an interest in maintaining and controlling their national borders as an aspect of their domestic, Westphalian, and interdependence sovereignty [6].

The widespread recognition of the state as a sovereign entity, in both domestic and international laws, reinforced the notion that human mobility should be subject to regulation, subordinating individual and collective will to state authority. As a result, migration ceased to depend solely on people's needs or desires and became subject to state-determined criteria of admission, exclusion, and management. This institutionalization of migration control, rooted in sovereignty, transformed mobility into a political instrument—a shift that demands an interdisciplinary perspective

[1]The year 1539 marks the posthumous publication of Francisco de Vitoria's *Theological Reflections*, most notably his *Relectio de Indis* (*On the American Indigenous Peoples*). In this seminal work, Vitoria posits that European people—specifically the Spanish—held a natural right to travel, engage in commerce, and establish residence within indigenous American territories, provided their actions caused no harm. This argument formed part of Vitoria's broader effort to reconcile colonial practices with moral and legal principles, framing such interactions within a framework of *ius gentium* (law of nations) and universal human dignity.

[2]The year 1625 marks the publication of Hugo Grotius's groundbreaking treatise *De Jure Belli ac Pacis* (*On the Law of War and Peace*), in which he systematically articulates principles of *ius gentium* (law of nations). Among these are the right to free passage, commerce, and colonization—concepts later instrumentalized to provide legal justification for European imperial expansion. Grotius further elaborates a naturalist framework for the *ius communicationis* (right of communication or transit), framing unrestricted global mobility as inherent to human sociability. He then extends this logic to legitimize the establishment of European settlements beyond continental borders, grounding colonial ventures in a universalist interpretation of natural law.

to analyze migration beyond traditional approaches focused solely on economics or demographics.

To navigate this complexity, the analysis must center on three key dimensions. First, the tension between state sovereignty and the right to regulate access to national territory must be considered. Second, and central to any ethical framework, the centrality of the human person and their fundamental rights—derived from human dignity, freedom, and the need for survival—must be prioritized. Third, there is an urgent need to promote peaceful coexistence among communities within the framework of a Global Bioethics grounded in solidarity, responsibility, and the construction of a system that fosters international peace and mitigates the exclusionary legacies of state-centric migration regimes.

Given the inherently transnational nature of migration—which simultaneously affects individuals, local communities, and global systems—addressing its challenges effectively requires a methodology capable of integrating these interrelated dimensions. Fragmented approaches, focused exclusively on state logic or administrative management of migratory flows, have proven insufficient in the face of such complexity. In this context, Global Bioethics emerges not merely as an academic discipline, but as a transformative lens that, by placing human dignity and collective responsibility at its core, enables the reconciliation of state sovereignty with human rights and communal coexistence. From this perspective, Global Bioethics offers holistic and sustainable solutions to the challenges of migration, oriented more toward long-term justice and equity than toward short-term control.

9.2 Contemporary Migration: A Brief Overview

Migration is a global phenomenon that has gained increasing complexity in the twenty-first century, driven by factors such as the climate crisis, geopolitical conflicts, economic inequalities, and the growing securitization of borders [7]. Unlike in previous historical periods, contemporary human mobility is shaped by a series of structural challenges that impact not only migrant populations but also countries of origin, transit, and destination. To unpack this complexity, the following discussion examines the principal legal classifications established under international migration law, emphasizing their key characteristics.

The analysis begins by assessing the frameworks devised by states and multilateral institutions to differentiate distinct migrant groups, as well as the rights associated with each classification. Crucially, these legal distinctions are not neutral; rather, they reflect state interests in regulating human movement and conditioning access to rights based on political priorities [8]. This instrumentalization of categories is evident in the way international law codifies migration and displacement through norms promoted by entities such as the United Nations (UN), the International Organization for Migration (IOM), and the Office of the United Nations High Commissioner for Refugees (UNHCR). Drawing on the expertise of scholars and migration professionals, these institutions have formalized the following typologies:

First, the most common category is that of the international migrant, defined as a person who moves from one country to another with the intention of residing temporarily or permanently [9]. This type of migration may be driven by economic, familial, educational, or other personal reasons. Their status is typically governed by domestic immigration laws, as well as international frameworks such as the *Universal Declaration of Human Rights* (Article 13) and the *International Convention on the Protection of the Rights of All Migrant Workers and Members of Their Families* (1990). These instruments, among others,[3] aim to safeguard the dignity of migrants, regardless of their administrative status in the host country.

Second, the category of refugee refers to individuals who have crossed an international border due to a well-founded fear of persecution based on race, religion, nationality, membership in a particular social group, or political opinion. This status is particularly critical, as such persons are unable or unwilling to seek protection from their country of origin, often fleeing life-threatening conditions [10]. The primary legal frameworks safeguarding this group include the 1951 Convention Relating to the Status of Refugees and its 1967 Protocol, which codify the principle of non-refoulement (prohibiting forced return to persecution). Closely linked is the right to asylum, which entails requesting protection from a foreign state before crossing its border, subject to the host country's legal procedures [11].

A third and less recognized category is that of *internally displaced persons (IDPs)*—individuals who, while remaining within their country's borders, are forced to flee their homes due to armed conflicts, human rights violations, natural disasters, or severe economic crises [12]. Unlike refugees, IDPs do not cross international borders, yet they experience similar losses of home and livelihoods. Their rights are recognized in the *United Nations Guiding Principles on Internal Displacement* (1998), a nonbinding framework that in practice often provides more limited protections than the constitutional guarantees afforded to a country's nationals. This disparity necessitates a comparative analysis of both protection systems to determine which offers greater safeguards for affected populations.

The stateless individuals constitute a fourth category in migration typologies—individuals without recognition as nationals by any state under its legal framework. This condition creates a protection gap distinct from other migratory statuses, as the absence of nationality automatically restricts access to fundamental rights and state protections. While international human rights law theoretically safeguards all persons, stateless individuals frequently fall through jurisdictional cracks, denied basic entitlements to education, healthcare, formal employment, and political participation [13]. Two UN conventions address this issue: the 1954 Statelessness Convention

---

[3] Other key international frameworks include the International Covenant on Economic, Social, and Cultural Rights (1966), which recognizes migrants' rights to work, health, and education (Art. 2.2, 6–13), and the Global Compact for Safe, Orderly, and Regular Migration (2018), a nonbinding UN agreement promoting cooperation on labor rights and human trafficking prevention.

(establishing minimum treatment standards) and the 1961 Convention on the Reduction of Statelessness (addressing root causes through nationality laws).

A fifth and increasingly critical category is that of climate-induced migrants—individuals forced to move due to environmental disruptions such as rising sea levels, droughts, floods, or extreme weather events. These situations are part of a set of emerging forms of forced displacement, which are becoming more frequent and severe as climate change intensifies. In many cases, entire communities are no longer able to remain in their places of origin due to the gradual collapse of ecosystems, water scarcity, or the devastation caused by hurricanes or wildfires. This has led to the growing use of the term "climate refugee" [14], highlighting the need to recognize these individuals as subjects of protection, similar to those fleeing war or persecution. However, unlike refugees under the 1951 Convention, climate-displaced persons do not yet have a legal status that guarantees them international protection, despite facing equally existential threats. Initiatives such as the Nansen Initiative have acknowledged this gap, but the lack of binding international frameworks leaves them in a state of legal and humanitarian vulnerability—especially as cross-border displacement is expected to rise in the coming decades. This legal vacuum exposes the limitations of current, state-centered protection systems, which are not designed to respond to the complex drivers of twenty-first-century displacement.

All these categories are grounded in the Westphalian model of Sovereign Statehood, whose internal legal framework—particularly administrative law—unilaterally determines conditions for entry, residence, and employment, prioritizing national interests such as labor demands in strategic sectors (agriculture, industry, etc.). However, this approach has become anachronistic in light of the evolution of international human rights law, which transcends the logic of absolute sovereignty by recognizing universal protections based on human dignity [15]. Those who fail to meet state criteria are labeled as irregular migrants, a designation that, while administratively convenient, perpetuates an obsolete legal paradigm by reducing individuals to binary categories ("legal" vs. "illegal"), disregarding states' extraterritorial obligations.

This critique is supported by the Inter-American Court of Human Rights in its Advisory Opinion OC-18/03 [16], where it states that all people—regardless of whether they have legal documents or not—must be treated equally and protected in their basic rights. The Court clearly affirms that no country can deny or limit someone's rights just because of their migratory status, since this would go against the basic principle that all human beings deserve equal treatment (para. 103).

The Court also explains that calling people "illegal" just because they don't have the proper documents is not only inaccurate, but harmful. These labels increase their vulnerability and contribute to discrimination and abuse (para. 11). Instead, countries must ensure that all people under their authority are protected, and that their actions respect human dignity at all times.

Finally, the Court warns that governments cannot use migration policies as an excuse to ignore or weaken the protection of human rights. Even if a country wants to control who enters or works there, this control must always respect people's fundamental rights, especially those related to work, safety, and justice (paras. 165–168).

Herein arises a normative tension: while the Westphalian model insists on state discretion, contemporary international law—from the 1948 Universal Declaration to the 2018 Global Compact for Migration—mandates the guarantee of basic rights (healthcare, education, due process, etc.) irrespective of migratory status. Thus, labeling a person as "illegal" not only undermines their intrinsic dignity but also contradicts the principle of progressive realization of human rights. Consequently, any State that denies these rights to migrants—irrespective of their legal status—not only violates international law and its ethical responsibility to protect human dignity but also perpetuates an obsolete Westphalian paradigm incompatible with twenty-first-century legal cosmopolitanism [17].

The categories described reflect the situation of need in which irregular migrants find themselves; however, force migrants appear to be in the greatest vulnerability, as their displacement often results in the loss of state protection and community support, making them susceptible to migrant smuggling—cross-border movement facilitated by traffickers in exchange for payment, usually without direct exploitation—and human trafficking, involving coercion, deception, or violence for purposes such as sexual or labor exploitation, forced begging, organ trafficking, or modern slavery. The relevant international frameworks include the UN Protocol to Prevent, Suppress, and Punish Trafficking in Persons, Especially Women and Children (2000) and the Protocol against the Smuggling of Migrants by Land, Sea, and Air (2000), both under the Palermo Convention.

However, despite the recognition of these categories and the legal advances surrounding them, significant practical challenges remain. While international law has developed various treaties and frameworks to protect migrants and displaced persons—ensuring access to fundamental rights and protections—contemporary challenges persist, such as the criminalization of migration, rising xenophobia, and insufficient international cooperation, all of which compromise the safety and well-being of millions on the move. Moreover, migration is a multifaceted issue that extends beyond migrant–state dynamics and increasingly generates conflicts within host communities, especially when there are no effective integration policies in place. The following section explores the principal challenges faced by irregular migrants and their impacts on receiving societies.

One of the main societal challenges concerning irregular migration is the structural vulnerability in which these populations often find themselves. Irregular migrants face significant barriers to accessing basic rights such as healthcare, education, housing, and decent work. In many countries, these essential services are limited to individuals with legal documentation, which forces irregular migrants to depend on informal networks or humanitarian organizations. This exclusion not only deepens their precariousness but also hinders their ability to integrate into host societies. For instance, in both the United States and the European Union, irregular migrants have restricted access to healthcare and education, reinforcing cycles of marginalization and social invisibility [18].

This lack of access to fundamental rights is closely linked to another major challenge faced by irregular migrants: labor exploitation and their dependence on the informal economy. Deprived of legal status, many are forced to accept exploitative working conditions, including extremely low wages, excessive hours, and the absence of labor protections. Their inability to report abuse or seek legal redress perpetuates illicit labor practices, particularly in sectors such as agriculture, construction, and domestic work [19]. In Spain, for example, irregular migrants represent a substantial share of the agricultural workforce, especially in fruit harvesting, where they often endure precarious conditions and work without formal contracts or social security coverage [20].

Closely related to these dynamics of exclusion is the rise of xenophobia and social conflict in host societies, as the arrival of irregular migrants can trigger negative perceptions among receiving communities, particularly when migration is associated with job competition, pressure on public services, or perceived insecurity. In many contexts, these fears are amplified by political discourse and media narratives that frame migrants as threats to national stability. In countries such as Italy and France, the rise of far-right movements has been strongly linked to the instrumentalization of migration issues, often blaming migrants for economic crises and unemployment [21].

Another critical dimension of irregular migration relates to the strain placed on public services and local resources in host countries. When migration occurs at high volumes without adequate policy planning or institutional support, essential systems such as healthcare, education, and housing may become overwhelmed. This often leads to social tensions, particularly in communities already affected by poverty, unemployment, or weak infrastructure. In Mexico, for instance, the recent increase in arrivals from Venezuela, Haiti, and Central America has severely strained migrant shelters and public services in border cities like Tijuana and Tapachula, exacerbating problems of marginalization and lack of integration [22].

Moreover, irregular migrants face significant challenges in integrating into society due to discrimination, language barriers, lack of documentation, and fear of deportation. These obstacles often result in the formation of urban ghettos and segregated communities marked by entrenched poverty and social exclusion, which heighten the risk of criminalization or exploitation by trafficking networks. In cities such as Paris and Los Angeles, many migrant communities reside in marginalized neighborhoods with low educational attainment, high unemployment, and limited access to opportunities for social and economic advancement, conditions that perpetuate cycles of exclusion and vulnerability [23].

Compounding these structural challenges is the constant risk of deportation and family separation faced by irregular migrants. The threat of detention and expulsion not only undermines their ability to build stable and secure lives but also generates deep emotional and psychological distress, especially when it leads to the separation of families [24]. In the United States, immigration enforcement policies have resulted in mass deportations that have torn apart family units, leaving thousands of U.S.-born children separated from their undocumented parents. These practices reinforce cycles of vulnerability and trauma, particularly among migrant communities already affected by poverty, discrimination, and legal exclusion [25].

In conclusion, irregular migration presents complex challenges for both migrants and receiving societies. While migrants contribute labor and cultural diversity, the absence of comprehensive regulation and integration policies fosters labor exploitation, discrimination, social conflict, and pressure on public services. Addressing these issues requires not merely the criminalization of migration, but rather the implementation of policies aimed at regularization, access to basic rights, and effective integration into host communities.

However, beyond legal reforms, there is a pressing need for ethical frameworks—rooted in global bioethics—that can guide and strengthen the international normative order. The persistent vulnerability of irregular migrants is not solely a legal failure but also a reflection of the absence of shared ethical principles capable of shaping inclusive, humane, and coherent migration policies. Without this ethical foundation, legal norms remain fragile and often ineffective in transforming the structural conditions of inequality and exclusion. This persistent disconnect between ethics and law has allowed the vulnerability of migrants to endure, highlighting the urgent need for a normative transformation that places human dignity at the center of global migration governance [26].

9.2.1 Migration Through the Lens of Global Bioethics

Building on the previous legal and social analysis, which was necessary to frame the problem, this section proposes Global Bioethics as a valuable framework for understanding human and social phenomena from a broader and deeper perspective. Moving beyond a purely normative lens requires us to question the foundational assumptions upon which modern statehood, economic systems, and social structures have been built, assumptions that, despite their variations, continue to shape contemporary institutions and public policies. In this regard, Global Bioethics offers a particularly useful approach. As a young and inherently interdisciplinary field, it draws from multiple sources—philosophical, scientific, cultural, and legal—to develop honest and nuanced understandings of persistent and increasingly complex issues, such as migration.

Global Bioethics, like Bioethics more generally, is grounded in ethical principles—understood as fundamental criteria that guide human conduct and decision-making in relation to what is considered right, just, or morally acceptable. These principles serve as the foundation for ethical reflection, critical analysis, and the development of alternative approaches in complex contexts such as migration. This section explores four key principles: pro persona, proportionality, necessity, and responsibility. Among them, the pro persona principle will serve as the central interpretive lens, given its normative relevance in prioritizing the protection of human dignity when legal norms come into tension with fundamental rights. From this perspective, the other principles will be examined not as abstract notions, but as practical tools for evaluating state actions and public policy. In this framework, particular attention will also be given to a critical discussion on state sovereignty and the ethics of migration control, aiming to question the legitimacy of restrictive migration policies when they conflict with ethical imperatives and universal human rights obligations.

9.2.2 The Pro Persona Principle

The *pro persona principle* [27] is a fundamental hermeneutical criterion in the interpretation and application of human rights norms, both in international law and in several domestic legal systems. Its core purpose is clear: when multiple interpretations are possible—whether of a legal norm or in resolving a conflict—the interpretation that provides the greatest protection for human dignity must prevail [28]. In this regard, the principle acts as a safeguard mechanism for all parameters related to the protection of human dignity, ensuring that any restriction is minimal, reasonable, and proportionate to the legitimate objective[4] pursued by the State or any other regulatory authority [29].

Within the field of global bioethics, the *pro persona* principle gains particular relevance, as ethical dilemmas in this field often involve tensions between legal norms, policy priorities, and the rights of vulnerable individuals. In this context, the principle guides decision-making by prioritizing human rights over institutional, political, or economic interests, ensuring that outcomes reflect a commitment to maximizing the protection of human dignity and minimizing unjustified restrictions.

A relevant example of the *pro persona* principle in global bioethics is the case of migration. This phenomenon raises ethical and legal tensions between the rights of migrants, the sovereign powers of States, and the concerns of host communities. In such contexts, the *pro persona* principle demands a careful examination of the rights at stake, emphasizing the dignity and well-being of individuals as a normative priority, without disregarding the regulatory role of States or the complex social dynamics present in receiving societies. This ethical focus is especially urgent given that dominant perspectives in migration ethics have often reproduced state-centered biases that prioritize national origin and institutional interests over universal standards of justice and equal moral consideration [30].

[4] A *legitimate objective* refers to a purpose that is recognized by law and ethically justified in democratic societies. It must pursue a socially valid aim—such as public security, health, or the protection of others' rights, while respecting fundamental rights and human dignity. In ethical terms, a legitimate objective cannot be reduced to institutional convenience or political interest; it must align with shared moral values that prioritize the well-being of individuals. In the context of migration, for example, a state may invoke national security as a legitimate aim, but applying this in ways that lead to arbitrary detention or family separation undermines proportionality and fails to meet ethical standards grounded in human dignity. Measures must be the least restrictive possible, and always aimed at protecting people, not merely regulating them [41].

As previously discussed, migrants—especially those in irregular situations—face multiple challenges that undermine their dignity and fundamental rights. From a bioethical and human rights perspective, human mobility should be recognized as an expression of personal freedom, the right to self-determination, and the pursuit of a dignified life. Freedom of movement and the expectation of being treated with dignity in the receiving country are essential dimensions of this experience, as they directly affect a person's identity, self-worth, and ability to care for themselves and others. These elements are fundamental to protecting individuals from situations of humiliation, oppression, and discrimination [31]. Yet in practice, migrants are frequently criminalized, discriminated against, and subjected to legal constraints that place them in conditions of extreme vulnerability.

Furthermore, in contexts of humanitarian crises and forced displacement, many migrants flee from conflict, extreme poverty, or environmental disasters in search of a life marked by dignity and security. However, host countries often adopt restrictive policies that limit access to asylum procedures, formal employment, healthcare, and education. The lack of legal recognition not only denies migrants access to basic protections but also exposes them to labor exploitation, human trafficking, and institutional forms of violence that directly undermine their rights and human dignity. In some cases, arbitrary detention or deportation can result in family separation, psychosocial trauma, and a forced return to environments marked by persecution, instability, or extreme deprivation—practices that contravene fundamental principles of human rights.[5]

In accordance with the *pro persona* principle, any migration policy or state action must guarantee the highest possible level of protection for human dignity, ensuring that no restrictive measure excludes or criminalizes migrants without a justification that is both ethically and legally sound, grounded in principles of proportionality and necessity. This imperative is not only juridical but moral, and it requires a profound reorientation of migration governance toward frameworks that prioritize human lives over institutional convenience or political calculation.

Beyond its legal relevance, the *pro persona* principle expresses a profound ethical conviction: that the person—especially those who are most exposed, displaced, and silenced—must always come first. In the context of global bioethics, this means recognizing that dignity is not an abstract value, but something lived and felt through the ability to be safe, to care for others, to build a future, and to be treated with respect. Migration challenges us to choose whether we will respond with control and exclusion or with responsibility and care. The *pro persona* principle reminds us that no system, border, or policy should ever come before the dignity of a human life.

---

[5] These patterns of restriction and harm have been widely documented by United Nations agencies in various reports addressing forced migration, protection gaps, and state practices. A recent example is the UNHCR's *Global Trends Report 2023*, which details the vulnerabilities faced by displaced populations and the ethical and legal challenges posed by restrictive migration policies [42].

9.2.3 State Sovereignty and the Ethics of Migration Control

On the opposite side of the balance, States possess the sovereign right to regulate the entry and residence of individuals within their territory. This prerogative is recognized under international law and is commonly justified on the grounds of national security, economic stability, or social welfare. However, sovereignty is not an absolute license: its exercise must be reconciled with binding international human rights obligations and subjected to ethical scrutiny, particularly when it directly affects the lives, dignity, and futures of vulnerable individuals. In a global bioethical framework, how sovereignty is exercised matters, not just in legal terms, but in moral terms, because it shapes who is protected, who is excluded, and under what justification.[6]

With respect to protection from other States, national sovereignty is grounded in the duty to safeguard citizens, which often entails the implementation of migration controls, entry requirements, and regulatory frameworks for managing population flows. Governments frequently invoke public security concerns to justify restrictions, arguing that large-scale arrivals may increase crime or overwhelm healthcare and welfare infrastructures. Economic arguments are also deployed to justify limitations on migration, particularly to protect domestic labor markets from perceived competition. While such concerns may appear legitimate within state-centric frameworks [32], they often obscure the ethical cost of exclusion and the human consequences of policies driven by fear, protectionism, or utilitarian calculations.

The ethical dilemma emerges when the exercise of State sovereignty crosses the threshold of respect for human dignity, giving rise to policies that systematically discriminate against, exclude, or criminalize migrants. In such instances, the *pro persona* principle acts as a normative limit to sovereignty, requiring that all migration-related measures be proportionate, demonstrably justified, and designed to uphold—rather than undermine—the fundamental rights of people in mobility. When this standard is not met, sovereignty ceases to be a protective function and becomes a tool of exclusion.

Host communities also experience tensions and challenges as a result of migrant arrivals. While migration can bring cultural and economic benefits, there are legitimate concerns that should be considered. In contexts of limited resources, local populations may feel that their access to healthcare, education, and welfare is affected by the arrival of migrants. Similarly, migrant labor can impact local labor markets by placing pressure on wages and working conditions, leading to resistance among some segments of the local workforce. Cultural identity is also a point of concern, especially in societies with strong national or cultural identities, where the presence of migrants may be perceived as a threat to traditional values, customs, or

[6] Based on the *Robust Respect Imperative*, it has been argued that the legitimate exercise of sovereignty in the context of migration requires state institutions to be structurally oriented toward systematically prioritizing the basic human rights of migrants. Where this ethical condition is not met, sovereign authority lacks moral legitimacy to enforce exclusionary migration policies [43].

languages, fostering xenophobic or discriminatory tensions. Finally, the perception of insecurity is often amplified by political and media narratives that associate migration with crime, feeding fears that may fuel hostile or exclusionary attitudes.

From the perspective of global bioethics, these concerns must be addressed without resorting to exclusionary or xenophobic rhetoric. Instead, they demand the creation of inclusive strategies that respect the rights of migrants while also attending to the legitimate needs of host populations. In this context, the principle of cooperation—as articulated in both the UNESCO Universal Declaration on Bioethics and Human Rights and in international legal frameworks—offers a valuable foundation for addressing such tensions. It emphasizes the ethical imperative to work together across borders and cultures to uphold human dignity, share responsibility, and build responses that do not pit one group against another, but instead seek justice and care for all affected communities.

9.2.4 *Pro Persona Principle and the Criteria of Proportionality and Necessity*

As discussed above, the *pro persona* principle is a central pillar in guiding decisions that impact human rights and bioethical dilemmas. Its ethical force lies not only in prioritizing the protection of individuals but also in demanding that any measure affecting human dignity be subject to rigorous evaluation. This includes the integration of the criteria of proportionality and necessity, which serve as safeguards to ensure that restrictions on rights are not only legally grounded but ethically justifiable, legitimate in purpose, reasonable in scope, and applied only when strictly required.

Resolving the tensions between the rights of migrants, state sovereignty, and the concerns of host communities requires an analytical framework that places human dignity at its core. In this regard, global bioethics provides essential tools for evaluating whether migration-related policies meet the ethical and legal standards required in democratic societies. The principles of proportionality and necessity are central to this assessment: they demand that any restriction on rights be fully justified, carefully calibrated, and applied only as a last resort. By foregrounding these criteria, the bioethical approach seeks to prevent the adoption of excessive, arbitrary, or discriminatory measures—and to ensure that public responses to migration remain consistent with both international law and the moral imperative to protect the most vulnerable.

9.2.4.1 The Proportionality Principle in the Restriction of Rights

The principle of proportionality [33] asserts that any restriction imposed on individual rights must be assessed considering its impact on human dignity and the legitimacy of its underlying objective. This means that adopted measures must not cause more harm than the problem they intend to resolve, and that they represent the least intrusive means available to achieve a justifiable aim. This principle is especially relevant in the context of migration, where regulatory policies may affect fundamental rights such as the right to life, personal liberty, or physical and psychological integrity. As such, proportionality operates as both a legal and an ethical safeguard against excessive, punitive, or unjustified responses to complex human mobility.

Therefore, a proportional approach demands that states, when regulating migration, adopt measures that maximize the protection of human rights without imposing disproportionate restrictions. For instance, prolonged detention of migrants in inadequate facilities, forced family separation, and deportation without risk assessment constitute actions that, under proportionality analysis, would be deemed excessive and unjustified. By contrast, establishing regularization procedures, reception mechanisms, and special protections for vulnerable individuals exemplifies how states can balance their sovereign prerogatives with their ethical and legal duties to respect, protect, and fulfill the fundamental rights of migrants.

The principle of proportionality must also guide how states respond to the concerns of host communities. While it is legitimate for local populations to demand safety and fair access to public services, any restriction placed on migrants must be grounded in objective criteria, not in fear, misinformation, or exclusionary rhetoric. Policies such as denying access to healthcare and education or criminalizing irregular migration often generate more harm than they prevent, exacerbating social fragmentation and deepening the vulnerability and marginalization of migrant populations. A proportional approach seeks not to choose between communities, but to foster solutions that uphold dignity and fairness for all [34].

9.2.4.2 The Criterion of Necessity in Migration Regulation

The criterion of necessity [35] holds that a restriction on human rights can only be considered legitimate if it is indispensable and if no less invasive or harmful alternative exists to achieve the same objective. In bioethical terms, this means that any measure limiting an individual's freedom, security, or well-being must be fully justified and represent the only viable option within the legal and ethical framework. Necessity, therefore, acts as a safeguard against arbitrary or excessive responses, compelling states to choose the least harmful course of action when regulating migration.

In the context of migration, the necessity criterion demands that state policies be evaluated according to their indispensability and proportionality. For instance, if a government seeks to justify the deportation of migrants on the grounds of protecting the local labor market, it must provide clear evidence that no less invasive alternatives—such as regulated labor integration or the development of equitable employment programs—are available. If such options exist and are not considered, the deportation measure would fail the necessity test and could be considered not only arbitrary or discriminatory but also ethically indefensible.

The principle of necessity also applies to the responses of host communities. While it is legitimate for local populations to raise concerns about the sustainability of public services or public safety, any restrictions imposed on migrants must be shown to be truly indispensable. Numerous studies have demonstrated that migrants do not represent a disproportionate burden on healthcare or social welfare systems;[7] in fact, their inclusion often contributes positively to economic growth and labor market vitality.[8] For this reason, it is ethically imperative to subject any restrictive measure to critical scrutiny—ensuring that it does not become a vehicle for unjustified exclusion or a response driven by prejudice rather than evidence.

In this sense, the principle of necessity requires that migration-related measures be grounded in reliable evidence, rather than in political rhetoric or unfounded perceptions. Any limitation on migrants' rights must be demonstrably indispensable, not a reflection of fear, prejudice, or xenophobia. When necessity is invoked without substantive justification, the result is not responsible governance but the erosion of human dignity and the distortion of public policy.

For all these reasons, the *pro persona* principle can be considered a cornerstone of global bioethics in addressing the ethical and normative dilemmas surrounding migration. Its application demands that public policies, legal frameworks, and judicial decisions be grounded in a criterion of humanity, ensuring that any restriction on migrants is proportional, justified, and truly necessary. Bioethical analysis based on these two principles offers a rigorous framework for establishing limits on state action and for safeguarding the rights and dignity of those affected by mobility. In this vein, proportionality ensures that restrictions are not excessive in relation to the legitimate aim pursued, while necessity obliges states to exhaust less invasive alternatives before resorting to measures that limit fundamental rights [36].

9.2.5 The Principle of Responsibility in a Global and Unequal World

The principle of responsibility, articulated by Hans Jonas in his seminal work *The Imperative of Responsibility: In Search of an Ethics for the Technological Age (1979)* [37], argues that ethics must integrate foresight and a duty to prevent harm—extending moral concern to both present and future generations. When applied to the context of global migration, this principle suggests that states and international actors cannot limit themselves to managing migratory flows; they must also acknowledge their role in producing the structural conditions that force millions to

[7] For example, a study in the United States has found that immigrants, including those without legal status, use significantly fewer healthcare services and incur lower medical costs than native-born citizens—contradicting common assumptions that migrants overburden public health systems [44].

[8] According to the OECD, migrants make a growing contribution to economic growth and labor market vitality, with rising levels of employment, innovation, and job creation across receiving countries [45].

migrate. As such, responsibility implies a dual obligation: to address the consequences of displacement and to actively transform the global inequalities and systemic failures that sustain it.

This perspective is essential in a world shaped by deep historical inequalities and enduring power asymmetries. Migration dynamics cannot be fully understood without examining the long-term effects of resource extraction, colonial domination, and political interference by hegemonic nations in the Global South [38]. Contemporary migration is not a spontaneous or isolated occurrence, but rather the result of centuries of unequal economic and political relationships rooted in dispossession, external intervention, and systemic destabilization. Recognizing this history is key to understanding why responsibility must go beyond border control and encompass structural transformation.

Applied to migration, the principle of responsibility calls for a comprehensive and ethically grounded approach to human mobility, one that recognizes the structural and historical roots of displacement. States and international institutions cannot confine their actions to border enforcement or the criminalization of migrants; they must confront their role in sustaining the global conditions that drive migration and adopt concrete, transformative measures aimed at justice, repair, and long-term solidarity.

This entails a transformation of global migration policies, prioritizing access to safe and legal pathways for displaced persons and promoting equitable development strategies in countries of origin. It also requires a reevaluation of economic and political relationships that perpetuate global inequality, ensuring that historically exploited nations gain sovereign control over their resources and acquire the tools necessary for autonomous development.

Moreover, ethical responsibility does not rest solely with institutions, it also extends to societies and individuals. Citizens of host countries share a moral obligation to reject xenophobia, foster intercultural dialogue, and support policies that uphold the dignity and rights of migrants. In everyday life, this responsibility may take the form of community solidarity initiatives, inclusive education efforts, or public discourse that challenges harmful stereotypes and cultivates empathy. A just and humane response to migration requires that individuals—especially those in privileged societies—recognize their entanglement in global systems that sustain inequality and displacement and act accordingly.[9]

Ultimately, migration is not an isolated phenomenon but a manifestation of longstanding structural injustices that demand responses rooted in justice, equity, and reparation, with human dignity as their ethical foundation. When aligned with the *pro persona* principle, the principle of responsibility in global bioethics reminds us

[9] In her influential work, Seyla Benhabib defends the ethical obligation of democratic societies to extend moral and political consideration to non-citizens, arguing that the legitimacy of democratic norms depends on their capacity to include the voice and dignity of "the other." She insists that individuals affected by the consequences of legal and political decisions—regardless of their status—must be granted the right to justification, as a core demand of moral universalism and democratic inclusion [45].

that addressing human mobility requires more than managing flows: it requires transforming the conditions that displace, exclude, and dehumanize [39].

9.3 Conclusion

Migration, more than a political issue or a legal challenge, is above all a profoundly human experience, one that reflects both our vulnerabilities and our shared hopes. In a world marked by deep inequalities and complex interdependence, addressing migration ethically requires more than technical solutions or reactive policies. This chapter has argued that Global Bioethics offers a powerful framework to rethink how we understand and respond to human mobility. Grounded in the principles of human dignity, responsibility, and care, it provides not only theoretical tools but also a sense of direction for shaping more just and compassionate societies.

By exploring key principles such as *pro persona, proportionality, necessity, and responsibility*, we have seen how Global Bioethics can help illuminate the ethical blind spots in current migration policies. The *pro persona* principle, for instance, reminds us that when laws or institutions come into conflict with people's dignity, it is the person who must come first. The criteria of proportionality and necessity call on states to act with restraint, fairness, and genuine justification, never out of fear or convenience. And the principle of responsibility urges us to go deeper, to recognize the historical and structural roots of forced migration, and to take seriously our collective duty to repair what has been broken.

What Global Bioethics offers, above all, is a different way of seeing. It invites us to move beyond the language of crisis and control, and toward an ethics of solidarity, one that sees migrants not as problems to be managed, but as people to be welcomed, protected, and understood. It challenges us to ask what kind of societies we want to be and what values we are willing to defend across borders and generations.

In the face of growing walls and shrinking compassion, this ethical perspective is not a luxury, it is a necessity. Global Bioethics reminds us that the defense of dignity does not stop at the edge of a territory. It begins wherever a person's rights are at risk, and it extends as far as our sense of justice, solidarity, and responsibility can reach.

References

1. Manning P. Migration in world history. In: Heikkilä E, editor. Which direction is Finland evolving? The dynamics of mobility and migration. Migration Institute of Finland; 2020. p. 18–39.
2. Buisel M. Troyanos y Latinos: migración e identidad. In: Cardigni J, et al., editors. Migraciones, desplazamientos, conflictos en el mundo antiguo. Buenos Aires: Editorial de la Facultad de Filosofía y Letras – Universidad de Buenos Aires; 2024. p. 131–43.

3. Stelder M. The colonial difference in Hugo Grotius: rational man, slavery and Indigenous dispossession. Postcolonial Stud. 2022;25(4):563–81.
4. Mahmud T. Cheaper than a slave: indentured labor, colonialism and capitalism. Whittier Law Rev. 1999;34:215–82.
5. Bhambra G. Colonial global economy: towards a theoretical reorientation of political economy. University of Sussex; 2021.
6. Torpey J. Coming and going: on the state monopolization of the legitimate "means of movement". Sociol Theory. 1997;16(3):239–59.
7. Mavroudi E, Nagel C. Global migration: patterns, processes and politics. Routledge; 2023.
8. Meyer CK, Boll S. Categorising migrants: standards, complexities, and politics. Anti-Traffick Rev. 2018;(11):1–14.
9. Nematov E. United Nations [Internet]. 2018 [cited 2025 Mar 31]. Available from: https://www.un.org/en/global-issues/migration.
10. Human Rights Education Associates (HREA). Study guide: the rights of refugees [Internet]. 2003 [cited 2024 Mar 31]. Available from: hrlibrary.umn.edu.
11. European Union Agency for Fundamental Rights. Article 18 – right to asylum [Internet]. 2007 [cited 2024 Mar 31]. Available from: https://fra.europa.eu/en/eu-charter/article/18-right-asylum.
12. High Commissioner for Human Rights. About internally displaced persons [Internet]. [cited 2024 Mar 31]. Available from: https://www.ohchr.org/en/special-procedures/sr-internally-displaced-persons/about-internally-displaced-persons.
13. United Nations High Commissioner for Refugees. Stateless person definition [Internet]. 2025 [cited 2025 Mar 31]. Available from: https://emergency.unhcr.org/protection/legal-framework/stateless-person-definition.
14. Apap J, Haarju S. The concept of 'climate refugee': towards a possible definition. Report No.: PE 698.753. Brussels: European Parliamentary Research Service; 2023.
15. Engle E. The transformation of the international legal system: the post-Westphalian legal order. Quinnipiac Law Rev. 2004;23(23):23–45.
16. Humanos CIdH. Condición jurídica y derechos de los migrantes indocumentados (Opinión Consultiva OC-18/03). 17 Sept de 2003. Serie A No. 18. Available from: https://www.corteidh.or.cr/docs/opiniones/seriea_18_ing.pdf.
17. Berebon C, Eluke P. Cosmopolitanism, sovereignty and global justice in light of refugee protection. GNOSI Interdiscip J Hum Theory Praxis. 2024;7(2):177–98.
18. Willen SS. Migration, "illegality," and health: mapping embodied vulnerability and debating health-related deservingness. Soc Sci Med. 2012;74(6):805–11.
19. European Union Agency for Fundamental Rights. Protecting migrant workers from exploitation in the EU: boosting workplace inspections. Vienna: FRA; 2018.
20. López-Sala A. Migrant labour and precarious work in Spain: the role of immigration policy. Report of the special rapporteur on the human rights of migrants on labour exploitation of migrants. 2020;Report No.8 (2).
21. Wodak R. The politics of fear: what right-wing populist discourses mean. World migration report 2022. London: IOM – International Organization for Migration; 2021.
22. Human Rights Watch. Mexico: events of 2023 [Internet]. 2024 [cited 2025 Mar 31]. Available from: https://www.hrw.org/world-report/2024/country-chapters/mexico.
23. American Civil Liberties Union. Family separation by the numbers [Internet]. 2022 [cited 2025 Mar 31]. Available from: https://www.aclu.org/news/immigrants-rights/family-separation-by-the-numbers.
24. Human Rights Watch. "We can't help you here": US returns of asylum seekers to Mexico [Internet]. 2021 [cited 2025 Mar 31]. Available from: https://www.hrw.org/report/2021/10/21/we-cant-help-you-here/us-returns-asylum-seekers-mexico.
25. UNHCR. Children on the run: unaccompanied children leaving Central America and Mexico and the need for international protection [Internet]. 2014 [cited 2025 Mar 31]. Available from: https://www.unhcr.org/children-on-the-run.html.

26. Boswell C, Geddes A, Scholten P. The ethics of migration policy dilemmas. Migr Stud. 2022;10(3):427–46.
27. Carbonell M. El principio pro persona en el Derecho Internacional de los Derechos Humanos. Rev Mex Derecho Const. 2010;(22):39–60.
28. Inter-American Court of Human Rights. Case of the "Street Children" (Villagrán-Morales et al. v. Guatemala). Judgment of November 19, 1999.
29. Corte Interamericana de Derechos Humanos. Opinión Consultiva OC-21/14. Derechos y garantías de niñas y niños en el contexto de la migración y/o en necesidad de protección internacional. 2014.
30. Dumitru S. The ethics of immigration: how biased is the field? Migr Stud. 2023;11(1):1–22.
31. Daly E, May JR, McClain SN, Bruch C, Hamada Y, Teji G, Maekawa M, Yoshioka N. Migration with dignity: a conceptual and policy framework. Geneva: IOM; 2021.
32. Dadhania PR. Reimagining sovereignty to protect migrants. Yale J Int Law Online. 2022;47:71–9.
33. OHCHR. The principle of proportionality under international human rights law. Geneva: United Nations; 2021.
34. Flynn M. Immigration detention and proportionality. Working Paper No. 4. Geneva: Global Detention Project; 2011.
35. Sanchís LP. Proporcionalidad y ponderación en el ámbito de los derechos fundamentales. Doxa. 2003;26:273–89.
36. Carens JH. The ethics of immigration. Oxford: Oxford University Press; 2013.
37. Jonas H. The imperative of responsibility. In: Search of an ethics for the technological age. Chicago: University of Chicago Press; 1984.
38. Bauman Z. Wasted lives: modernity and its outcasts. Cambridge: Polity Press; 2004.
39. Sassen S. Expulsions: brutality and complexity in the global economy. Cambridge: Harvard University Press; 2014.
40. Office of the United Nations High Commissioner for Human Rights (OHCHR). Recommended principles and guidelines on human rights at international borders. United Nations Office of the High Commissioner for Human Rights; 2014.
41. United Nations High Commissioner for Refugees (UNHCR). Global trends: forced displacement in 2023. Copenhagen: United Nations High Commissioner for Refugees; 2023.
42. Schmid L. Saving migrants' basic human rights from sovereign rule. Am Polit Sci Rev. 2022;116(3):954–67.
43. Kaiser Family Foundation. Key facts on health care use and costs among immigrants. San Francisco: KFF; 2023.
44. OECD. International migration outlook 2024. Paris: OECD; 2024.
45. Benhabib S. The rights of others: aliens, residents, and citizens. Cambridge: Cambridge University Press; 2004.

Chapter 10
Sustainable Development Goals and the Universal Declaration on Bioethics and Human Rights

Henk ten Have

10.1 Introduction

In September 2015, all member states of the United Nations adopted the *2030 Agenda for Sustainable Development* with the aim to promote peace and prosperity for all people on the planet as well as for future generations. Its core comprises 17 Sustainable Development Goals (SDGs) providing specific goals for a practical framework of international cooperation and action. For the next 15 years, these goals will guide the international community in addressing global health, biodiversity loss, and social inequalities. Ten years earlier, in October 2005, all member states of the United Nations Educational, Scientific, and Cultural Organization (UNESCO) adopted the *Universal Declaration on Bioethics and Human Rights*, formulating ethical principles in regard to medicine, life sciences, and associated technologies. Although focused on a specific field of action (i.e., medicine and healthcare), the Universal Declaration is addressing similar global challenges as the SDGs but emphasizes in particular the ethical principles that should motivate international action when faced with these challenges. This chapter will discuss the interrelations between these two international documents.

H. ten Have (✉)
Center for Global Health Ethics, Duquesne University, Pittsburgh, USA

Department of Bioethics, Anahuac University, Mexico
e-mail: tenhaveh@duq.edu

D. Cerdio-Domínguez, M. E. de los Ríos-Uriarte (eds.), *Current Bioethics*, New Paradigms in Healthcare, https://doi.org/10.1007/978-3-032-25294-4_10

10.2 The Sustainable Development Goals

The SDGs are considered as global goals that should transform the world. It is clear that agreement on these goals presupposes that there is consensus about what are the serious and widespread problems that are threatening humanity now and in the future. This consensus has emerged in a long process of international deliberation and policy recommendation. The first steps were taken at the Earth Summit in Rio de Janeiro in 1992 with the adoption of *Agenda 21* [1]. In order to achieve sustainable development, it was necessary to combat extreme poverty especially in developing countries, promote health, and protect the environment (fight against deforestation, conserve biodiversity, control pollution, and manage biotechnology). The action plan acknowledged that economic growth and increasing globalization have produced growing inequalities, fragilized health and healthcare, and deteriorating environments across the world. The benefits of growth and globalization have produced unequal outcomes especially in developing countries, and international action is deemed necessary to reduce such inequalities.

An important milestone was the *Millennium Declaration* adopted at the Millennium Summit in 2000 in New York, which elaborated the Millennium Development Goals (MDGs). In this document, countries went one step further in formulating specific targets to be achieved in 2015 [2]. It clearly expressed that in addition to separate responsibilities of individual societies, countries have a collective responsibility to improve the living conditions for people and the planet. Eight goals were set. Some identified what are regarded as major challenges: eradicate of extreme poverty and hunger, reduce child mortality, improve maternal health, combat diseases such as HIV/AIDS and malaria, and ensure environmental sustainability. Other goals are more focused on means and ways to better address these challenges: universal primary education, gender equality and empowerment of women, and global partnerships for development.

In 2012, the member states of the United Nations started to work on the policy agenda after 2015. They adopted at the United Nations Conference on Sustainable Development in Rio de Janeiro the document *The Future We Want* [3]. The commitment to sustainable development is renewed, but it is also noticed that insufficient progress has been made. They decided to start an intergovernmental process to develop global sustainable development goals, striving for a world that is just, equitable, and inclusive. The following year, the General Assembly of the United Nations set up a working group to develop a proposal on the SDGs, which was finally adopted in 2015. The range of goals is wider than previously formulated [4].

- End poverty in all its forms everywhere.
- End hunger, achieve food security and improved nutrition, and promote sustainable agriculture.
- Ensure healthy lives and promote well-being for all at all ages.
- Ensure inclusive and equitable quality education and promote lifelong learning opportunities for all.
- Achieve gender equality and empower all women and girls.

- Ensure availability and sustainable management of water and sanitation for all.
- Ensure access to affordable, reliable, sustainable, and modern energy for all.
- Promote sustained, inclusive, and sustainable economic growth, full and productive employment, and decent work for all.
- Build resilient infrastructure, promote inclusive and sustainable industrialization, and foster innovation.
- Reduce inequality within and among countries.
- Make cities and human settlements inclusive, safe, resilient, and sustainable.
- Ensure sustainable consumption and production patterns.
- Take urgent action to combat climate change and its impacts by regulating emissions and promoting developments in renewable energy.
- Conserve and sustainably use the oceans, seas, and marine resources for sustainable development.
- Protect, restore, and promote sustainable use of terrestrial ecosystems, sustainably manage forests, combat desertification, halt and reverse land degradation, and halt biodiversity loss.
- Promote peaceful and inclusive societies for sustainable development, provide access to justice for all and build effective, accountable, and inclusive institutions at all levels.
- Strengthen the means of implementation and revitalize the global partnership for sustainable development.

This list illustrates not only that the goals are ambitious but also very disparate. Some are evidently directed at global challenges that are threatening human well-being, such as poverty (SDG 1), hunger (SDG 2), health (SDG 3), inequality (SDG 10), and climate change (SDG 13), while others are more related to economic factors, such as energy (SDG 7), work (SDG 8), industry (SDG 9), and consumption and production (SDG 10). There is not a specific or overarching goal of environmental protection, but there are specific goals related to the environment, i.e., clean water (SDG 6), life below water (SDG 14), and life on land (SDG 15). Concerns for biodiversity loss and planetary survival do not figure prominently. One difference with the MDGs is that the focus no longer primarily is on developing countries but now on all countries. Another difference is that for each goal, specific targets and indicators for outcomes and implementation are formulated so that progress can be measured and reviewed. Nonetheless, it is not likely that the goals will be met in 2030. The COVID-19 pandemic has provided a setback in the realization of goals, especially with inequalities increasing around the world, and biodiversity loss and climate change continuing. At the same time, it is significant that the world community has accepted this framework as expression of its aspirations that all people should live in peace and prosperity.

10.2.1 Search for Common Values

The 1990s were characterized by a search for shared values at the international level. The Commission on Global Governance, for example, convened in 1992 at the occasion of the 50th anniversary of the United Nations articulated the idea of "global neighborhood" [5]. All people in the world are in the same predicament; they all share the same neighborhood, which is the planet. Without an appropriate ethics to preserve and protect this neighborhood, the planet will not survive. Such ethics should express that humanity as a whole will be best served if it is recognized that there are common rights and responsibilities and core values that all humanity should uphold. In the view of the Commission, these values are: respect for life, liberty, justice and equity, mutual respect, caring, and integrity. A few years later, the United Nations' *Millennium Declaration* emphasized the notion of common humanity in all its diversity. In order to make globalization fully inclusive and equitable, the following fundamental values are regarded as essential to international relations: freedom, equality, solidarity, tolerance, respect for nature, and shared responsibility [6].

Similar searches for shared values were going on in other areas. In religion and theology, the activities of the Parliament of the World's Religions were aimed to show what world religions have in common rather than articulating how they differ. In 1993, more than 200 leaders from more than 40 religions and spiritual traditions launched the statement *Toward a Global Ethics*, drafted by the German theologian Hans Küng [7]. It declares that all traditions share common values such as respect for life, solidarity, tolerance, nonviolence, and equal rights. In the field of medicine and medical science, the Council for International Organizations of Medical Sciences (CIOMS) adopted the *Declaration of Ixtapa* in 1994 [8]. Discussing poverty, vulnerability, and the value of human life, the Declaration highlighted the need to develop a common framework of ethical principles to guide the development and application of medical sciences across the world. The fact that moral values and practices differ cannot be used as an excuse not to explore and identify fundamental ethical principles that are valid for all human beings and that should be applied everywhere on the globe. The call for a global agenda for bioethics was an important stimulus for subsequent UNESCO initiatives.

In hindsight, it is obvious that the era around the new millennium was entirely different from the present time, where differences rather than commonalities in international relations are celebrated (questioning whether, for example, even the United States and Europe share the same values), and where globalization is regarded as a problem rather than an opportunity. Current policies tend to blame processes of globalization for the ills of present-day societies (such as migration, inflation, unemployment, infectious diseases, and shortages of essential medication) and to retreat into nationalistic policies. The main argument is similar to the one that was frequently used during the COVID-19 pandemic to justify lack of international cooperation and solidarity: the first duty of governments is to protect their own citizens, even if that means that other citizens are harmed. At first sight, this seems a return to tribal ethics. Moral considerations are deemed to apply to members of one's own group or nation but not to strangers. A distinction is made between insiders and outsiders, and the horizon of ethics is not extended to all human beings. This point of view is justified with references to evolutionary theory: humans have an innate disposition to engage in cooperative behavior with members of their own group and not to take the perspective of out-group members [9]. Within-groups behavior is regulated by moral emotions such as empathy and gratitude facilitating reciprocal relations between group members while those outside the group are met with aggression and revenge. The ideas of self-interest, competition, and conflict are relocated to the intergroup level or as visible today to the interactions between nations. Taking the perspective of other groups or nations is regarded not only as contrary to the moral dispositions inherent in human beings but also as something that should not be done since it is threatening the flourishing and survival of one's own group or nation. Against this backdrop, relations between groups or nations are simply transactions and should only be undertaken if they serve the best interest of one's own group or nation. The atmosphere around the new millennium was quite different because the problems related to globalization were perceived from a cosmopolitan point of view. Two factors especially motivated the search for common values. One is the recognition of the global nature of the challenges that humanity is facing. Climate change, inequality, and poverty, for instance, cannot be addressed in a piecemeal manner by separate actors and from a national perspective but require cooperation. Second is the awareness that the capacity to deal with global problems is very unequally distributed within and even more between countries. While globalization is first of all benefitting richer nations, its disadvantages are mostly experienced by developing countries that have less resources to effectively deal with them. Because problems are global, their impact cannot be confined to a particular area or region of the world; ultimately, they will affect all countries, as has been illustrated in the COVID-19 pandemic. It will be in the interest of humanity as a whole that these problems will be addressed everywhere, and that requires international solidarity. These two factors remain relevant today, even when the search for common values is denied or rejected from the point of view that self-interest of nations has priority. It is a strong motivation to continue to articulate what humans have in common, and to pursue cooperative approaches to solve, or at least mitigate global challenges, as articulated in the Sustainable Development Goals.

The international documents discussed so far also point out characteristics that are important for the development of a global ethical framework. The first is that global problems are complex and interrelated. One problem cannot be solved without addressing other challenges. Poverty cannot be eradicated when health is not improved, and health cannot be improved as long as biodiversity is deteriorating. The 17 SDGs are therefore integrated. They recognize that action in one area will affect the outcomes in others. Development must find a balance between social, economic, and environmental sustainability. This refers to the second characteristic, which is particularly significant for bioethics: human health is not merely an individual concern but intimately connected with social and environmental concerns. That implies an important deviation from the perspective of bioethics as it has developed over the past few decades [10].

10.2.2 The Universal Declaration on Bioethics and Human Rights

In October 2005, the member states of UNESCO unanimously adopted the *Universal Declaration on Bioethics and Human Rights (UDBHR)* [11]. This was an unique event since for the first time in the history of bioethics, governments and the international community committed themselves to fundamental bioethical principles. While other international documents, most importantly the *Declaration of Helsinki*, play an influential role in medical research, they do not provide commitments of governments (since they are adopted by professional organizations but can then be incorporated into national or regional legislation). The UDBHR is a normative instrument with a global outreach and is part of international law (although not binding in distinction to a Convention).

That UNESCO (rather than WHO) has been the platform to develop the Declaration can be explained with references to its specific mission. UNESCO is the only United Nations organization concerned with culture, education, and science. It was founded in 1945 to create a genuine culture of peace through education, science, culture, and communication. The constitution of this new organization was to give hope to the international community for a world of solidarity and dignity after the world war that had just ended. It is the only United Nations (UN) agency with specific responsibilities for scientific research. This implies the promotion of scientific collaboration among the peoples of the world in order to advance the objectives of international peace and of the common welfare of humankind. It also implies making sure that the advances of science and technology will take place within the framework of "universal respect for justice, for the rule of law and for the human rights and fundamental freedoms," as expressed in Article 1 of UNESCO's constitution. The ethical aspects of scientific research therefore need to be considered. Science is not regarded as an end in itself but as a means to address the basic problems and needs of the world population and to resolve global problems such as poverty, environmental degradation, and child mortality. Science can help to implement fundamental values and should in its development and application take into account all perspectives that are relevant to all countries across the world.

With the emergence of the life sciences in the 1970s, and around the same time as the term "bioethics" has been coined, UNESCO started to organize activities in the field of bioethics, especially related to scientific and technological innovations in genetics, the life sciences, and reproductive technologies [12]. To facilitate reflection and activities, the International Bioethics Committee (IBC) was established in 1993 in connection to a work program and budget (with a separate administrative unit in the Social Sciences Sector). The main objective of UNESCO in bioethics is the development of international normative standards. Activities first focused on the human genome and the effort to draft an international instrument in this area. This resulted in the adoption in 1997 of the *Universal Declaration on the Human Genome and Human Rights* and in 2003 of the *International Declaration on Human Genetic Data.* These successes encouraged the launch of a bolder initiative in 1997: The

Universal Ethics Project, focused on the question whether it would be possible to formulate bioethical principles that can guide not merely genetics but also global bioethics in its entirety. The fact that European countries succeeded in adopting the *Oviedo Convention* (The Convention for the Protection of Human Rights and Dignity of the Human Being with regard to the Application of Biology and Medicine) in 1997 stimulated this search for global consensus and demonstrated the feasibility of connecting bioethics with international human rights.

According to Van Rensselaer Potter, who coined the term 'bioethics" in 1971 and expanded his vision with the concept of global bioethics in 1988, contemporary bioethics should be oriented toward the future; a bridge between the present and the future because for the survival of humankind, it is vital to focus on long-term interests and goals [13]. The ultimate goal of bioethics is long-term global human survival. For this purpose, new knowledge and old wisdom should be combined. Otherwise, the pressing global problems of today cannot be properly addressed. Another point is that the goal of human survival can only be accomplished when there is a balanced social and ecological system for human life to flourish. Bioethics therefore should do more than presenting an ethical framework focused on the individual good; it should also refer to the common good as an anchor for social justice and environmental sustainability.

The ethical principles declared in the *Universal Declaration on Bioethics and Human Rights* provide precisely such an expanded framework. The principles determine the different obligations and responsibilities of the moral subject (*moral agent*) in relation to different categories of moral objects (*moral patients*). The principles are arranged according to a gradual widening of the range of moral patients: the individual human being itself (human dignity and human rights; benefit and harm; consent; autonomy), other human beings (persons without the capacity to consent; respect for human vulnerability; privacy and confidentiality; equality, justice and equity), human communities (nondiscrimination and non-stigmatization; respect for cultural diversity and pluralism), humankind as a whole (solidarity and cooperation; social responsibility; sharing of benefits), and all living beings and their environment (protecting future generations; protection of the environment, the biosphere, and biodiversity). Although no hierarchy among principles is implied, this ordering means that the focus of ethical concern is not restricted to an individual point of view; ethical concern should equally take into account common social and ecological perspectives. Global bioethics therefore cannot be merely medical, social, or environmental; it should include all perspectives at the same time. At a global level, bioethics should have a wider focus and be concerned with humanity as a whole.

10.2.3 The Bridge Between Global and Local

Nowadays, globalization is increasingly criticized by people arguing that it should be rejected, restricted, or redefined. Experiences during the COVID-19 pandemic illustrated how global interconnectedness has weakened the abilities of countries to effectively respond to the viral threat. Since then, nationalistic discourses in various parts of the world have only become stronger. Against this backdrop, the framework of global bioethics has also become problematic. It is argued that the human condition is defined by moral diversity. Nations and populations have their own views on what is best to do. Since an overarching ethical worldview is impossible, there is a plurality of particular approaches to bioethics (such as Mediterranean bioethics, Chinese bioethics, Islamic bioethics, and Black bioethics) Another argument takes this view one step further: global bioethics is the imposition of a specific form of particular bioethics. Since bioethics has emerged in the West, declaring that it is global is in fact universalizing Western moral views and therefore a form of ethical colonization or imperialism [14].

Criticism of globalization and global ethics often assumes that global processes are more or less autonomous, imposing a specific moral order on the world that is erasing local differences and homogenizing various cultures and traditions, leading to a bioethical monoculture. While it is true that they proceed from a specific normative framework (which I will elaborate below), this view disregards the dialectics between the global and the local. Global ethics is a two-level phenomenon [15]. At the abstract level, there is the global ethical discourse defining a minimum set of standards agreeable to all. At the contextualized level, there is a multiplicity of different ethical traditions. These "local" traditions define what is ethically required beyond and above these principles. On the one hand, there is a set of standards on which traditions and cultures agree; this is elaborated into specific bioethics principles in connection to international human rights language. On the other hand, there are many efforts to articulate more specific bioethics standards in the context of particular religious and cultural traditions. Members of these traditions also bring their views in the global debate through constructive dialogues and sometimes negotiations, so that the dialectic of global and local also helps to construct, produce, and refine global bioethics. The universal principles of global bioethics are the result of continuous and multilateral articulation, deliberation, and production. This is exemplified in the *Universal Declaration on Bioethics and Human Rights.* The request to develop a common framework of ethical principles in bioethics was explicitly made by developing countries. They were afraid that with the rapid evolution and globalization of medical science and research, they would insufficiently benefit from the advances and suffer too many harms and risks. A major concern was that international medical research and healthcare endeavors would proceed along double standards so that people in developing countries would receive substandard care and be involved in clinical trials without the ethical protection that exists in developed countries. The adoption of the Declaration shows that agreement could be reached on 15 global bioethics principles. However, these universal principles need to be interpreted, implemented, and applied in specific local settings of different cultures and traditions. The abstract and contextualized levels are therefore interacting along bottom-up and top-down lines of communication. Global platforms and local contexts mutually help each other to construct and produce global bioethics. Thus, global bioethics is the result of continuous and multilateral articulation, deliberation, and production. While global and local levels can be distinguished as two levels of ethical discourse, they are not separated but entangled. The local and global, according to French philosopher Edgar Morin, are implicated in each other [16]. When the local is isolated from the whole of which it is a part, it becomes abstract. Likewise, as soon as the global is detached from its localized parts, it is converted into an abstract notion. Since global and local are dialogically and recursively related, there is no opposition between the two. Global policies do not imply that local approaches cannot be heterogeneous, while at the same time, local policies are contributing to global approaches. The same is true for people who are citizens of a particular nation while being citizens of the world; being cosmopolitan does not entail that we are rootless, abstract individuals.

10.3 A Different Mode of Globalization

Contesting global bioethics often presupposes a narrow or one-sided view of globalization. It is difficult to deny that human existence has significantly been altered over the past few decades. Even the most zealous opponents of globalization are using its accomplishments to criticize global processes with the help of computers, mobile telephones, internet, email, and social media. We all benefit from greater accessibility to information, from easier communication with Zoom or Teams, and from the products of international cooperation and research. At the same time, it has become clear that globalization has negative effects: increasing inequalities, growing vulnerabilities, environmental degradation, emerging infectious diseases, to name a few. It is undeniable that globalization has two faces: prosperity for a minority and continuing misery for large populations.

Nonetheless, it is doubtful whether globalization as such can be blamed for these outcomes; it seems rather that they are the adverse effects of how and in what way globalizing processes are taking place; in other words, that the negative outcomes are the result of a particular mode of globalization that has been dominant during the past decades. This mode is neoliberal market ideology that has principally determined global policies and practices. Neoliberalism is defined as "the set of discourses, practices and apparatuses that determine a new mode of government of human beings in accordance with the universal principle of competition." It is also a normative stance, "… a system of norms now profoundly inscribed in government practices, institutional policies and managerial styles" [17]. Neoliberal policies are multiplying insecurities: less and more precarious employment, deterioration of working conditions, financial instability, growth of poverty, and environmental degradation. They also lead to the breakdown of protective mechanisms; safety networks and solidarity arrangements that existed to protect vulnerable subjects have been minimized or eliminated. Rules and regulations protecting society as well as the environment are weakened in order to promote global market expansion. As a result, precariousness has generally expanded. This is precisely what the ideology wants to accomplish: people only flourish if they are confronted with challenges, if there is the possibility of competition. Individual security is "a matter of individual choice" [18]. The devastating effects of neoliberal policies are most visible in the developing world. But nowadays, existential insecurity is everywhere. It is also obvious that market ideology has not in fact increasing human welfare. It has mainly promoted increasing inequality. It has created a world in which the 85 richest persons have as many financial resources as the 3.5 billion poorest people [19]. A small elite has appropriated the political process and has bended the rules of the economic system for its own benefit. Read the story of Iceland: in the 1970s and 1980s, an egalitarian country with a rapidly growing economy. Neoliberal policies and privatization of the banking system in 1998–2003 resulted in fast enrichment of a small elite but massive indebtedness of the country so that in 2004, it had the highest national debt in the world [20].

Arguing that the roots of contemporary problems are in neoliberal policies and practices has consequences for the conception and methods of bioethics. When bioethics discourse was initiated and expanded during the 1970s and 1980s, the major moral challenges were related to the power of science and technology. How can patients be protected against medical interference and paternalism? How can citizens have more control over healthcare decisions? In what ways can patients' rights be defined and implemented? These questions have shaped the agenda and methodology of mainstream bioethics, especially in more developed countries. But in a global perspective, many citizens do not have access to modern science and technology. They are marginalized in a system that is increasingly privatized and commercialized. They are exploited in clinical research projects since it is their only change to receive treatment and care. It is obvious that in this perspective, especially since the 1990s, the major moral challenges have changed. It is no longer the power of science and technology that produces ethical problems but the power of market logics, the pursuit of profits and money. When all areas of society are deregulated and privatized, borders have become irrelevant, and healthcare, research, education, and even culture and religion are regarded as businesses that compete for consumers, bioethics is no longer confronted with usual challenges such as reproductive technologies, organ transplantation, abortion, and euthanasia, but also with a new type of challenges: brain drain, human trafficking, biopiracy, corruption, food insecurity, health tourism, disaster relief, lack of access to healthcare, and poverty.

When neoliberalism is the dominant rationality, there is no room for other ethical principles than the maximization of self-interest. Competition is the global norm. In some sense, the starting point is an apocalyptic scenario: the bureaucratic welfare state is destroying the virtues of civic society, and undermining individual values, taking away personal responsibility and dignity. At the same time, neoliberalism has created a new human condition governed by the rules of competition and efficiency and with society as an enterprise of enterprises. It has also liberated the individual from the constraints of governmental intervention. Human life is the result of private choices, and humans are not simply regarded as calculating beings maximizing benefits and minimizing costs but as competitors always working to produce a better situation for themselves [21]. Neoliberalism therefore displays a new rationality and, moreover, a normative system that is affecting all dimensions of human existence. The belief that market solutions have moral superiority crowds out every other moral consideration, and makes the formulation, let alone the application of other moral judgments futile.

When the sources of global bioethical problems are within the neoliberal logic of globalization rather than in globalization itself, and when this logic has become a pervasive normative system, bioethics can only operate within the bounds of this system. This is exactly what mainstream bioethics has been doing. It articulates, just like neoliberal policies, that the individual is an active, creative, and constructive rational subject that alone should decide and that determines what values are worthwhile and achievable. The efficient, competitive individual is best at managing his or her own life and constantly transforming himself or herself in a process of self-improvement. The world cannot be changed, but humans can master and invent themselves, hence the importance of ethical principles such as respect for individual autonomy, informed consent, ownership of the body, and personal responsibility. Human vulnerability in this perspective is defined as lack or deficiency of autonomy requiring protection and empowerment, but not as the result of socioeconomic conditions that make the existence of some people more precarious than that of others [22].

Global bioethics on the other hand provides a different ethical horizon. First of all, it attempts to transcend the context in which today's moral challenges arise and to undertake a critical examination of the normative neoliberal framework. As long as bioethics is captured in this framework, and articulates the same set of values, it will be impossible to properly address global moral problems. Second, it develops a broader ethical framework that brings into play not only values related to individual well-being but also values that highlight social well-being and environmental flourishing. The UNESCO Declaration provides such framework as a broader normative view for the world. Third, global bioethics advances not only diverse concepts and values but also different methods and practices. Since global problems are complex and interconnected, an interdisciplinary approach is required that is broader than mainstream bioethics, engaging not only life sciences, philosophy, and medical science and technology but also social sciences, theology, political sciences, economics, international law, global studies, and comparative anthropology. Because critical examination of the dominant normative framework will necessarily include critical analysis of power relations and differences, the usual distinction between bioethics and biopolitics is obliterated. Furthermore, practical activities are undertaken beyond the typical domain of academic research and education. More than in mainstream bioethics, it is unavoidable to engage in policymaking, public debate, advocacy, and social activism and to create alliances with multiple actors around the world.

10.4 The Challenge of Implementation

While the SDGs identify the major global problems and provide practical guidance on how to tackle them, the moral reasons why it is imperative to address these problems are not explicitly mentioned, although associated statements such as the *Millennium Declaration* refer to common values. The *Universal Declaration on Bioethics and Human Rights* provides a systematic overview of the significant values and principles in connection to health and healthcare. The listed principles can therefore be regarded as the ethical motivation and justification for the goals adopted by the world community as SDGs.

How can the global ethical framework help to implement the goals of sustainable development? Going back to what has been said earlier about the dialectics of global and local, two different perspectives on implementation should be distinguished [23]. The first is that the global standards adopted by the international community should be diffused and translated into practices through legislation and regulation. Implementation here is conceived as compliance. It is up to the states to apply the agreed standards in their territory. But in the current climate of deregulation and privatization, this approach to implementation is rather weak, even more so since implementation mechanisms and accountability at the global level are almost missing. The example of international human rights law is illustrative. Human rights treaties are ratified by most states, but noncompliance with treaty obligations is more common than compliance. However, this does not mean that nothing is done. Commitments as well as violations are monitored and publicized by international bodies and NGOs. The International Court of Justice can mediate in conflicts between states. The International Criminal Court can prosecute individuals for crimes against humanity and has done so in multiple cases, currently, for example, investigating human rights violations in the wars in Ukraine and in Palestine, issuing warrants of arrest for the Israeli Prime Minister and Minister of Defense [24]. The second perspective on implementation assumes intensive interaction between global and local values. Human activities do not simply change because principles are imposed or theories are applied but are transformed through reinterpretation and construction of transnational principles in local practices. Implementation requires "localization," which is a dynamic process of building convergence between global principles and local practices and is therefore decentralized and domestic. What is necessary is empowerment of domestic agents, involvement of grassroots movements and social networks, effective use of various media, as well as deliberation and participation of local stakeholders and civil society.

10.5 Conclusion

Globalization is not merely a set of processes encompassing and homogenizing the world, but it is the result of intensive dialectical interaction between the global and the local. Goals for sustainable development cannot simply be imposed from a global perspective, but their implementation demands a range of activities at local levels. The same is true for global bioethical principles and values; they need to be domesticated within regional and national systems and local contexts. Governments can be stimulating the relevant actors and creating the conditions to undertake appropriate actions, but usually, nongovernmental organizations and cooperating individuals carry out the daily tasks of achieving goals and targets through collective agency. This is exactly the purpose of the SDGs. Goals for sustainable development can be defined by intergovernmental organizations, but they encourage and inspire citizens, individual and collective, all over the world to undertake activities directed at achieving the goals. Global bioethics provides the framework to motivate and facilitate dialectical interaction, interpretation, internalization, and transformation of principles into practices. It presents a moral horizon that shifts the perspective from the individual person and national entities to the perspective of what is common and what is the destiny of humankind and its planetary home.

References

1. United Nations. Agenda 21. [Internet]. 1992. Available from: https://sustainabledevelopment.un.org/content/documents/Agenda21.pdf.
2. United Nations, General Assembly. United Nations millennium declaration. [Internet]. 2000. Available from: https://ww.un.org/en/development/desa/population/migration/generalassembly/docs/globalcompact/A_RES_55_2.pdf.
3. United Nations. The future we want. Rio+20. United Nations Conference on Sustainable Development. A/Conf.216/L.1*. [Internet]. 2012. Available from: https://wedocs.unep.org/bitstream/handle/20.500.11822/13662/N1238164.pdf?sequence=1&amp%3BisAllowed=.
4. United Nations development program. What are the sustainable development goals? [Internet]. 2025. Available from: https://www.undp.org/sustainable-development-goals.
5. Commission on Global Governance. Our global neighbourhood. Oxford, New York: Oxford University Press; 1995.
6. United Nations, General Assembly. United Nations Millennium Declaration. [Internet]. 2000. Available from: https://ww.un.org/en/development/desa/population/migration/generalassembly/docs/globalcompact/A_RES_55_2.pdf.
7. Parliament of the World's Religions. Toward a global ethics. Chicago: Council for the Parliament of the World's Religions; 1993.
8. CIOMS. A global agenda for bioethics: declaration of Ixtapa. Can J Med Technol. 1995;57:79–80.
9. Wright C. Tribal morality and the ethical other: the tension between modern moral aspirations and evolved moral dispositions. J Dialog Stud. 2015;3(1):89.
10. Ten Have H. Global bioethics. An introduction. London, New York: Routledge; 2016.
11. UNESCO. Universal declaration on bioethics and human rights. Paris: UNESCO; 2005.
12. Ten Have HAMJ, Jean MS, editors. The UNESCO universal declaration on bioethics and human rights. Background, principles and application. Paris: UNESCO Publishing; 2009.
13. Ten Have H. Potter's notion of bioethics. Kennedy Inst Ethics J. 2012;22(1):59–82.
14. Ten Have H. Global bioethics; 2016. 76 ff.
15. Kymlicka W. The globalization of ethics. In: Sullivan W, Kymlicka W, editors. The globalization of ethics. Religious and secular perspectives. New York: Cambridge University Press; 2007. p. 1–16.
16. Morin E. Homeland earth. A manifesto for the new millennium. Cresskill: Hampton Press; 1999.

17. Dardot P, Laval C. The new way of the world. On neo-liberal society. London, New York: Verso; 2017. p. 4–14.
18. Harvey D. A brief history of neoliberalism. Oxford/New York: Oxford University Press; 2005. p. 168.
19. OXFAM. Working for the few. 178 Oxfam briefing paper. [Internet]. 2014. Available from: http://www.oxfam.org/sites/www.oxfam.org/files/bp-working-for-few-political-capture-economic-inequality-200114-summ-en.pdf.
20. Reid S. Iceland's saga. Foreign Aff. 2014;93(1):142–50.
21. Dardot P, Laval C. The new way of the world. Verso Books; 2014. p. 256.
22. Ten Have H. Vulnerability as the antidote to neoliberalism in bioethics. Rev Redbioetica/UNESCO. 2014;5(1):87–92.
23. Ten Have H. Global bioethics; 2016. 188 ff.
24. International Criminal Court. Situation in the State of Palestine: ICC pre-trial chamber I rejects the State of Israel's challenges to jurisdiction and issues warrants of arrest for Benjamin Netanyahu and Yoav gallant. [Internet]. 2024. Available from: https://www.icc-cpi.int/news/situation-state-palestine-icc-pre-trial-chamber-i-rejects-state-israels-challenges.

Chapter 11
Bioethics and Armed Conflicts

Carlos Alfonso Viesca y Treviño and Maríablanca Ramos R. de Viesca

11.1 Introduction

Armed conflicts have been constant and lamentable situations throughout humanity's history. Wars, with their consecutive confrontations between soldiers, are not limited to armed conflicts but also produce diverse types of aggressions and abuses on civil populations, with rapes, genocide, the displacement of populations from their homes and lands, and slavery.

Also, as has been most frequent in recent times, internal armed conflicts have developed in many countries with guerrilla groups. These frequently constitute sudden and brief attacks on villages and urban suburbs but also include successful and established dictatorial governments involved in punitive and repressive conduct against anyone who is suspected of any kind of opposition or who disagrees.

In its origins, bioethics does not include these armed aggressions against human beings among its essential fields of knowledge and action. Van Rensselaer Potter expressed in his initial and substantial work his concerns about the indirect or unconsciously desired harmful effects of scientific and technological discoveries on the environment and human health, expressing severe doubts on the real benefits obtained through its everyday most intense diffusion and its inaccurate uses in many fields of civilized life. His utopian view of a wisdom acquisition inside an evolutionary way toward noosphere was a significant contribution, opening the field of bioethics and the start of many reflections and its consecutive actions directed toward creating a consciousness of the responsibility of each human and each human institutions to the fight against the always-present danger of disease, environmental

C. A. Viesca y Treviño (✉) · M. R. R. de Viesca
Department of History and Philosophy of Medicine, Faculty of Medicine, National Autonomous University of México (UNAM),
Mexico City, Mexico, Brasil 33, Ciudad de México 06020
e-mail: cviesca@frontstage.org

D. Cerdio-Domínguez, M. E. de los Ríos-Uriarte (eds.), *Current Bioethics*, New Paradigms in Healthcare, https://doi.org/10.1007/978-3-032-25294-4_11

exhaustion and fatal biosphere deterioration. But curiously, his view does not specify the awful effects of armed conflicts on human survival and well-being.

In this chapter, we will present some reflections on bioethics and medical attention in war, the problems related to civil populations, and the reality of genocide and armed conflicts in everyday life in official "peace" situations, all of which will be treated with argumentation on the imperative necessity of bioethical responses, not only concerning medical and health problems but also concerning political decisions at the global level.

11.2 Development

11.2.1 Bioethics and Medical Attention in War

Centuries before the birth of bioethics, medical practitioners and the people attending soldiers, such as chargers, cooks, women in domestic activities, and complete families in the case of invading armies, had long been delivering medical attention to soldiers. In this sense, the ethical compromise of physicians and surgeons has always been evident and well done in most cases. Their obvious ethical duty was always to pay attention to the people in their army. In this sense, enemies were outside of their professional ethical area of responsibility, as were conquered or invaded civil populations, who were not only left out of any class of care or medical attention but also made victims of every kind of abuse. The modification of the fellow-being concept proposed as an absolute modification of the Good Samaritan parable concerning the traditional way of considering as a potential enemy anyone from a different tribe or race, maintained by the Christian religion, was not taken into account in war situations or followed for soldiers professing other faiths or for Christian enemies.

One significant step arrived when armies acquired a major range of displaced people and could not transport injured soldiers with the advancing or retiring army. These wounded soldiers had to be left them abandoned on the battlefield or in its vicinity, with the immediate possibility of being prey to other enemies or animals. The first response was the invention of the pike dagger, which some distinguished knights and lansquenets possessed to kill their injured compatriots. This ethical resolution was a real case of euthanasia in the most precise sense of the word. This conduct appeared in the fifteenth century in response to the increased mobility of armies.

A little later, in the sixteenth century, some military surgeons adopted ethical positions concerning treating their compatriots and war prisoners. Ambrosio Paré, who became a personal surgeon to the French kings, is a perfect example. Paré remembers some specific situations and evokes his reflections about the surgeon's compromise in making surgical interventions on the battlefield or its immediate surroundings and taking care of wounded soldiers. Talking in the context of Turin military actions on the possibility of killing wounded soldiers, euthanasia, or compassionate killing, he expresses that it is not the correct one. Still, his reflection speaks of the priority of making an accurate prognosis and, in the case of having possibilities for an effective treatment, taking action immediately. His account is clear: Inside a stable where he and another man intended to leave their hoses, they found four corpses and three badly wounded soldiers completely disfigured and unconscious; asked by an old soldier if it was possible do something for them, Paré replied that their prognosis was fatal; told the old soldier "*doucement et sans colére*"; cut their throats, explaining to the surgeon that he hopes to have someone nearby who will make the same decision for him in a similar circumstance [1]. Interpreting this as a definitive refusal of pity killing, practiced by the surgeon or by a fellow soldier but always with a surgeon's advice, may be considered as plausible in cases without any survival possibility, but according to the Paré religious, Protestant convictions, the fact could be considered as an absolute negative, interpreted by Gross as something to "vigorously condemn" [2].

Many years later, in the last years of the eighteenth century, Jean Dominique Larrey, who some years later would be the chief surgeon of the Napoleonic Imperial Guard, was the devised two essential and definitive innovations: the introduction of ambulances and giving medical and surgical attention to injured soldiers on the field of battle, regardless of whether they were rival army soldiers or civilians. In his experiences published as *Clinique Chirurgical*, he recalls his participation in the transportation of casualties to hospitals, speaking on the urban revolt and street combats at Paris in July 1830, consisting of more than two hundred young citizens and civilians, in which he makes special mention of a case of an injured young pharmacist [3]. In another way, this time about the professional responsibility to attend to people, in particular drowned soldiers considered as irrecoverable, he described and called attention to the effective maneuvers for the expulsion of inspired water [3]. Two anecdotal issues in the Larrey biography are evidence of his committed involvement in the ethical obligation to attend to injured enemies on the battlefield. In the Rhin campaign, from 1792 to 1793, he cared for an Austrian prince and carried out an amputation on an injured soldier who left in liberty, a situation which was the pretext for a denunciation against him and a judicial process for treason in the Comité de Salut Public; Larrey recognized the facts, adducing that the injured and disabled man could not go back to the army, but the Revolutionary Tribunal supported condemning him to death. A National Convention decree, signed on May 23, 1793, established that "medical treatment will be obligatory and the same, in the military hospitals, for the enemy's soldiers and the national soldiers." Fortunately, Larrey was exonerated of the treason accusation through the intervention of military authorities [4, 5]. Also, at the end of the Battle of Waterloo, Larrey, a prisoner to be executed, was released by Prussian Field Marshal Blücher, because Blücher recognized him and remembered that Larrey, sometime ago, had attended to his son, then a severely injured prisoner of the French [4, 6].

The most crucial treatise about soldiers' health problems published in 1768, John Pringle's *Diseases of the Army*, features no mention of wound treatment or the difficulties or characteristics of medical attention in and around the campaign [7]. In all armed conflicts, battle injuries go hand in hand with a diversity of illnesses and other health problems derived not only from the environment but also from hunger, displacement, and military operations, imposed on civil populations—which still apply today.

These antecedents speak clearly about the importance of the conscience of some surgeons working on the battlefields and taking on a professional duty to assist any injured people—soldier or civilian, enemy or comrade in arms. But the incredible and sad thing is that, generally speaking, the care of wounded soldiers was left aside and not considered as an essential matter, except for some individual military surgeons' actions, as shown by Henri Dunant, after he was an eyewitness at the Battle of Solferino in 1859. This was followed by the installation of some services for attending to French wounded soldiers and Austrian wounded soldiers and by later decisions in 1864 found in the Geneva Conference, which led to the foundation of the International Red Cross and to Dunant's receiving the first Nobel Peace Prize [5, 8, 9].

Medical care for war victims has been progressively improving from World War II up to today, and physicians involved in their attention have, at least in theory, more and more-efficient resources. But the possibility of accessing it is only one part of medical attention and, for our actual purpose, of medical bioethics, which involves many additional considerations, such as the primary defense of the vulnerability of soldiers but also, and no less important, that of civil populations, and the maintenance of a paternalistic attitude confronting the impossibility or at least diminished possibility of making decisions for affected people. The ability to transport seriously injured patients for immediate air transport to third-level hospitals is an incredible achievement and represents a bioethical distributive-justice dilemma.

11.2.2 The Role of Medicine in Armed Conflicts

The first moral dilemma derives from considering whether the ethics of medicine and the ethics of war are or at least could be the same. The 2004 *Regulations in Times of Armed Conflicts* of the World Medical Association is definitive: "medical ethics in times of armed conflict is identical to medical ethics in times of peace." Gross, in the first pages of his *Bioethics and Armed Conflict*, cites a Pellegrino and Thomasma consideration about the physician's obligations, which "transcend self-interest, exigency and even social, political and economic forces"; "physician's moral obligations remain absolutely constant"; and "War transforms, contracts, and subordinates the traditional doctor–patient relationship but rarely preserves it intact as in peace time" [2, 10, 11]. Finally, beneficence toward and respect for people are clues for the bioethical exercises of medicine. Thus, the main questions are, what transformations does medical ethics undergo during armed conflicts, and do the ethical obligations of physicians, surgeons, and other healthcare professionals remain the same when facing different types of patient identities? Also, we must consider whether human rights are truly universal or could be abolished or at least diminished depending on the identity of each individual's role in specific moments.

The first fundamental consideration, whose validity has been contested, as we have seen from times before ours, is whether healthcare professionals should give medical attention to soldiers of an opposite army. This consideration needs to factor in the character of healthcare professionals, who would be, by definition, noncombatants. It also needs to factor in the historical evidence of the double role that military physicians needed to assume when they were sometimes fighting enemies in defense of the hospital or the provisional place organized for the injured attention and at other times binding the same adversary's wounds [12]. However, most military actions did not consider or recognize this separation during the nineteenth and twentieth centuries. The main obstacle has been that medical personnel are officially included in the armed forces and have a military degree or, at least, its equivalent. Therefore, they have obligations stemming from the military's code of ethics. The evident problem is the possible opposition between their medical professional duties and their military duties. In any specific situation, does military necessity supersede any other ethical consideration? Is the imposition of criteria from military necessity or a matter of state? There is evident opposition in medicine in war between using medical resources to maintain one's fighting strength and using them to carry out the principal objective of medicine, medical attention for all who need it [2, 9]. Medical research in wartime and in the intermediate times between armed conflict and the preparation for upcoming ones has distinctive characteristics similar to those of research on biological weapons, the development of the COVID-19 virus, and studies on the radiation effects of nuclear weapons [9, 13, 14]. Is working on these research projects ethical? Is doing so moral, and if so, from which point of view? Is possible national utility a correct argument for the rupture of the main bioethical principles of the researcher? The other side of the question is the possibility of obtaining valuable resources for medical or environmental benefit, in which case this research will be justified. But what always emerges is the fact that the research projects, at their origins, were not intended for medical advances but for weapons development. Then, we must first insist that—in a utopian view of an ethically congruent world—professional health researchers not accept participating in these particular research projects. This is, of course, partly utopian. However, it is also part of a fervent desire for an improved world without wars or colonialism, a bioethical assertion of the evolutionary path for humankind toward the noosphere.

Also, the grave conflict between national war interests and the international obligation to prevent war or military actions in one's own country cannot be ignored. This conflict forces physicians and healthcare providers to face a personal and deep bioethical dilemma, which may be irresolvable. The measure of bioethical activism has limits, sometimes between a personal ethical dimension and the common good and sometimes between a well-founded desire and actual reality.

Searching for a familiar agreement to accept reasonable rules for protecting vulnerable people, some international agreements were made in the late nineteenth century. The Hague and the Geneva Convention, which were adopted in 1949 and have been maintained as valid until today, have been invaluable elements that have inspired public policies. The expressed areas to regulate were understood in the Convention for the Amelioration of the Wounded, Sick, and Shipwrecked Members of Armed Forces at Sea, the Convention Relative to the Treatment of Prisoners of War, and the Convention Relative to the Protection of Civilian Persons in Times of War, this last one prohibiting deportation, hostage taking, torture, and discrimination. Since then and until today, however, it has suffered severe ruptures on each one of its components in declared and undeclared wars and internal armed conflicts in diverse nations and has suffered the repression of different kinds, all of them affecting medical actions. An important factor is also the variation of rights depending on patient identity. In the fight for the establishment and global recognition of human rights, the validity of the Universal Declaration of Human Rights by United Nations Educational, Scientific, and Cultural Organization (UNESCO) is limited and often left out of consideration in times of war. Human rights differ between friendly soldiers and enemy soldiers, between combatants and noncombatants, between civilians and soldiers, and between different civilian groups depending on proximity to and sympathy for the given army. They determine rights; the conditions of invasion or military occupation; the recognition of racial, religious; or political differences, and, In internal conflicts the existence of irreductible factions. The rights applied to soldiers as patients have a definite scope derived from the fact that they are, by definition, people limited in their autonomy and, consequently, in making their own decisions. For them, medical care acquires a reinforced paternalistic dimension. Then, the right to informed consent for any medical procedure involves factors of military discipline and increases risks to compatriots; refusing treatment, as an example, could reduce the number of active soldiers but could also put them at risk if the military authorities decide to use their soldiers to test unproven chemical agents. As an example of this last situation, the increasing frequency of exposing soldiers to biological and chemical weapons, searching for the development of immunity, involves the risk of resulting damage from the harmful agents against the possibility of becoming resistant to them. In extreme military risk situations, the opportunity to refuse treatment or oppose any order from their military superiors is nonexistent and does not require minimal informed consent [2].

Confidentiality and the maintenance of professional secrets in times of armed conflict are also complicated. An infection should not be spread, because it represents a risk for the community in encampments and shelters. However, for other noncontagious problems, healthcare professionals' attitude is essential to safeguard the patient's privacy and reputation. Some regulations oriented toward humanitarian obligations specify the following obligation: "not to divulge any information that may be harmful to [a physician], patient or the latter's relatives" [2]. This obligation is valid for victims in noninternational armed conflicts with an explicit consideration of the family and friends, which, without even implying potential dangers to society, could be prejudicial for them if disclosed. Generally speaking, the exercise of professional confidentiality by physicians and healthcare personnel is a prima facie moral obligation; it comes, not in wartime, from the Hippocratic oath and seeks to protect the patient from any harm caused by revealing private information. But maintaining it implies a conscious responsibility for the healthcare professional, who needs not only to practice good judgment to respect and protect the patient's privacy but also in armed conflict times not to disclose information that could result in danger or damage to the healthcare professional's compatriots. As Gross expresses, medical information is a patient's property. It is an essential component of their exercise of autonomy and, in this sense, is a high priority of bioethical medical practice [2].

Another concept full of doubts and controversies is the right to die. Profound differences emerge between rights in wartime and rights in peacetime, and one of the most evident ones is the right to die, which implies the form and conditions of death. This right implies euthanasia, active or passive. One direct application of this possibility, converted into a right, is euthanasia on the battlefield. History, as we have seen in the first pages of this chapter, tells us that in desperate conditions, dying is better than being injured and unable to flee or act in one's minimal defense. It opens the possibility of suicide, and the legalization of suicide in some countries has made it more accessible, including (medical) assistance in dying, which is ultimately a disguised euthanasia, a pitiable one, as Kevorkian's support insists through a lot of personal awful experiences in judicial processes and injuries coming from public opinion [15]. The practice of the right to die for critically ill or injured soldiers has some interesting facets. The first is the impossibility of offering them hope for treatment on the battlefield or in its surroundings. This situation is becoming less probable in actual cases of battle-injured soldiers who have a fatal prognosis. A second one is the change in status of the soldier, who has become unable to return to their military activities and is converted into a nonmilitary person; they then recover the status of a civilian patient and the right to decide, without military-defined impositions, whether they want continue with or refuse treatment, confronting the possibility of never going back to their previous military duties. They should then have the right to decide as any other civilian patient would, a situation that usually is not recognized by military authorities.

Piet's perspective on killing out of compassion has changed substantially in modern warfare. If before the problem derived from the difficulty of their transportation in a forced withdrawal, contemporary technological possibilities have made it less probable but not impossible. Abandoning sick or wounded soldiers implies the risk that they fall into the enemy's hands and have a high possibility of being tortured, mutilated, and executed. Now, the decision made by officers, or mainly by the army commandant, is to search for a less prejudicial resolution—to kill sick or wounded soldiers to minimize suffering instead of leaving them abandoned to their inevitable horrible fate. These decisions have been called subsidiary to the doctrine of double effect because of the supposed benefit derived from an evidently prejudicial act, such as killing them. The transition from the right to die to the right to kill wounded soldiers in the name of military necessity is an actualized version of the ancient practice of killing in action on the battlefield. Still, its implementation brings another problematic aspect: Who would be an adequate person to kill a compatriot out of compassion [2]? Historical records feature many experimental soldiers. In modern times, the action falls back more and more into the hands of surgeons or military physicians because the means that they employ are not throat cutting but rather the injection of lethal substances accompanied by other drugs with anesthetic or sedation properties. These sad but often-inevitable circumstances bring to us another major ethical dilemma: the ethics of killing in any type of armed conflict, involving defined or ambiguous enemies, terrorists, civilian populations in various dimensions, and even genocide. We will come back to discuss this bioethical problem in the following pages.

Another bioethical dilemma is derived directly from triage, which in peacetime operates under the reason of a more equitable distribution of scarce resources and best possibilities to recover health, but never leaves out anyone from potentially being treated. This is the orientation held by the Geneva Convention, but North Atlantic Treaty Organization (NATO) operational guides, for example, give priority to military recuperation. The most frequent medical decision is war is the selection of who is to be attended to first, which depends not on the severity of patients' health conditions but rather on prioritizing the patient to be treated while leaving aside the urgency of the case. Patients in war, generally speaking, represent various possibilities; they are not only patients but also soldiers or civilians who could be friends or enemies, who could physically or emotionally recover and thus fight again and who could be capable of or suspected of being capable of engaging in armed resistance against the physician's army. In this way, selection first prioritizes friendly soldiers before enemies and second prioritizes combatants before noncombatants. Among friendly soldiers, those with a significant probability of returning to military duties should be selected first for their survival, but always prefer combatants before noncombatants. The triage selection of patients to attend to operates on a "salvage" criterion, which prioritizes giving attention to those who are "immediately useful" [2, 16].

From a bioethical point of view, a severe conflict emerges between so-called war ethics and medical ethics in their actual dimension as medical bioethics. The person's identity as a human being, owner of a project of life, is relegated to taking into account only their role in the panorama of the armed conflict in which they are living at this precise moment. People are valued according to their roles. This is an evident diminution in the value of ethics, in this case bioethics. Bioethical principles were left aside; they were only secondary to military objectives. Autonomy is minimized; beneficence is conceived only in terms of the utility of the army in question, and the same can be said of justice and nonmaleficence. The protection of vulnerable people is a secondary aim. In this way, a conflict emerges in the definition of a good physician, which in the war ethics world is a professional modified to be a relatively nonautonomous subject, with diminished autonomy that privileges military duties and their superiors' orders. In this context, a physician's bioethical dimension is subordinated to conditions defined in war actions and circumstances. In Pellegrino's defined search for virtue with beneficence as a primary goal, this situation leaves the physician limited in their capability to act according to medical decisions, leaving aside many possibilities to benefit patients, but it may convert them into virtuous military physicians.

In this sense, Gross's criticism of the utilitarian ethical conception of war medical ethics is, from our point of view, precise and just. Utility will be considered a limiting fact, not a value. Defined as limiting harm and achieving good, it deserves some consideration [2]. From the original utilitarian concept, the benefits of war ethics for a numerical majority of people or of nations with their inhabitants could be completely justified. What is the objective of cost–benefit analysis? In hard-to-define benefits for an immense number of people, atomic bombs exploded in Hiroshima and Nagasaki could be, and were, justified by the United States and Allied military and political directives, considering that a lot of lives would be lost if the war continued, that those lives were saved, and that the victims of the explosion and those of the subsequent effects of radiation were less important or valuable. Is this a utilitarian justification? It is an invalid justification, independent of numerical calculation. The proposal of a principle of proportionality, limiting the so-called military necessity and avoiding *excessive* harm, opens up possibilities to humanitarian concern for human lives in general and offers a reasonable bioethical solution for many cases of armed conflict [2, 17].

What is the real value of a human life? Is living a human life with dignity, naturally subject to individual values and expectations, adequate? All these considerations become obstacles in the context of war ethics. In a medical context, bioethics is a common problem that needs to be solved.

All these considerations lead us to reflect on dignity. The word's origin takes us to ancient Rome, where the insignias conferred on the people elected for a public charge converted them into *dignitario*, those who become possessors of dignity and public recognition. Centuries later, at the end of the fifteenth century, Giovanni Pico della Mirandola published the *Oratio de hominis dignitate* (Oration on the Dignity of Humankind), supporting the Christian perspective that all humans possess an intrinsic dignity derived from "God's image and similarity." This is only the starting point. Human dignity, considered as an essential characteristic of all humans, has been increasing in value, becoming a source of bioethical reflection and the basis for bioethical actions to respect the individual dimensions of each of us, promote their existential fulfillment, and protect them when in vulnerable conditions, such as illness and injury. Human dignity is an intrinsic and absolute value that implies that every human is entitled to all human rights and liberties; its essence refers to value, quality, and status, and consequently, everyone deserves estimation and honor [18]. In this sense, two philosophical antecedents arise: John Locke assumes that dignity is the result of the integral action of human rationality, and David Hume, in a more concrete way, assumes it as a result of putting into practice the ensemble of human moral ideas. The inclusion of morality as an essential component of dignity constitutes a fundamental element of the conception of the human person as an ethical being and not only a person recognized socially as a function of their access to public services. An interesting consideration, from Michael L. Gross, is that which puts self-esteem as a first-level definitional characteristic of dignity as a result of a conceptualization of dignity as a principle and not as a qualitative property of human beings, an intrinsic quality of them, and in this way, self-esteem is a source of moral and bioethics principles. Healthcare professionals are to respect self-esteem and avoid humiliating a patient—or a human being subdued to any other in circumstances of necessity, as in disease or subordination to an army organization; these are direct consequences not only of compassion, charity, and honoring professional obligations but of recognizing the integral humanity of the other, seeing them as another self and, consequently, see in them our self-esteem with all the different qualities defining a rational and moral being, our dignity. In this sense, dignity is an absolute value [2].

In the bioethical realm, human dignity has acquired an obligatory and coactive dimension, but always while respecting and promoting personal decisions, not only in individual situations but also in any contingencies involving bioethical dilemmas [19]. From this, dignity needs to be considered as a polysemic concept, which has been subject to changes according to circumstances and the philosophical bases of different thinkers, but when we speak exclusively of human dignity, the concept reaches a nonvariable and nonmodifiable minimum that is absolute for all humans, consistent in the refusal of any discriminative approach and the recognition of their intrinsic and irreducible value [20]. This property does not admit degrees, nor is it susceptible to being annulled by anyone or in any circumstances, and when elevated to the quality of a primary human right, it is to be respected as universally obligatory [21]. This has a natural consequence, as a property of the body, not only on organ donation and assisted fecundity [22] but also, which is the case of our interest here, on decisions and conduct in armed actions that do not follow the "normal" aggression toward enemies and the noncombatants of an opposing group [23].

Some later reflections left the field open for future consideration. Examples of these come from Patrice Churchland, from her neurophilosophical perspective, positing that if the brain makes decisions, humans and our dignity could lose their value or our sense of existence, but in her conclusions, human dignity is "a worthy aim" [24]. Furthermore, Martha Nussbaum emphasizes the substantial importance of dignity as a limiting factor against or in favor of political decisions and impositions [25].

Without controversy, human dignity as an absolute and unquestionable right is a fundamental bioethical component in the fight against all types of injustices and for supporting autonomy, including in authority or military situations, as a primordial beneficence vehicle [26, 27]. Hanna Arendt is an exemplary representative of the promotion of human dignity as an ineluctable factor for establishing now-recognized human rights politics [28]. In Gross's opinion, dignity during armed conflicts "functions as an overarching moral principle that yields several subsidiary principles that we term *human rights*" [2]. Also, perhaps the most representative of the vital mission of global bioethics, not just medical bioethics, is Henk ten Have's contribution, insisting on its substantial role in developing a deep consciousness of the obligation to protect human vulnerability in every possible circumstance [29].

Thus, the necessity to respect and protect vulnerability has been reinforced and has become an element in the development of conscience about the responsibility inherent in any situation of possible power abuse, besides vulnerable people. In this sense, respect for and the protection of dignity are primary bioethical obligations, which in armed conflict times remain in force and define how actions should be carried out or modified. Their natural obligation, derived from the duty of care of a military physician and any other healthcare professional or nonprofessional incidentally charged with these functions toward caring for injured soldiers, is to give injured soldiers the best management and preferential treatment. History has given us many examples of the quality surgical treatment of wounded soldiers and their efficient evacuation from the battlefield, by using, for example, airplane ambulances for evacuation in the Iraq War and the Gulf War. We also have testimonies of heavy casualties, like those suffered by Mao's army during the 1950s campaign, that were relatively resolved via traditional medicine and local resources. In all cases, with efficient and deficient medical services, the first question is whether governments and other leaders are aware of their responsibility and bioethical obligations and invest all their most or less limited resources into medical services for wounded or diseased soldiers, taking as a referent the high costs of healthcare versus the possibilities of injured soldiers' returning to their previous military functions. Is applying crude utilitarianism in economic terms morally justifiable from a bioethical viewpoint? The moral effects on troops from knowing that they will be medically well attended to is important in the bioethical concern of their superiors; in a reasonable utilitarian view, exalts their moral compromise; and signifies the accomplishment of the bioethical duties of superiors in the army and political leaders.

Recently, in bioethical reflections, the right to life for military personnel has been questioned in light of the fact that all of them are highly exposed to death. Also, they certainly lose the common civil rights and liberties at the precise moment of their enlistment or their access to a military-defined profession, including that of military physicians. The conjunction between medical necessity and military necessity is an ideal objective during armed conflicts, often conducive to heroic attitudes and morally correct actions, but it also compels physicians and attendants to face dilemmas when caring for enemies—soldiers and civilians—highlighting their neutrality in moral and bioethical terms in ignoring the war's concerns to eliminate enemies or at least leave them to perish, which in reality is the correct fulfillment of their professional ethical duties [2, 9, 30]. One relevant problem is the imposition on physicians by military authorities to act in a way absolute contrary to that established by their professional ethics, where so-called malevolent practices contribute to aggression, hostility, interrogations, and torture [30, 31]. The literature gives us some illustrative examples: Pearl S. Buck's *The Enemy* provides a sensitive narrative of the care given by a physician to enemy where the physician prioritizes their professional ethical duty; Thomas Mann's *Confessions of Felix Krull* features a physician who is deeply critical attitude of the ruling social morals of the German Nazi state and the war ethics in all activities, civil and military; and Izzeldin Abuelaish's autobiographic narrative features a physician who continues working, even in Israel, and caring without discriminating against Jewish patients or Palestinian patients instead of adopting a vindictive attitude against Israel after the violent death of his two daughters in an Israeli bombardment on civilian infrastructure in Gaza [32].

Aside from giving medical attention to injured soldiers and civilians always and most frequently and in an organized way, humanitarian organizations have been working very effectively in the aspects here involved, in a genuine, ethical way. Médecins Sans Frontières (Doctors Without Borders), groups organized by UNESCO, and the recently British teams of physicians organized in London and working in Gaza are some examples. Humanitarian ethics has opened the field not only to war but also to taking as their standard the ethical obligation to aid in any disaster, such as earthquakes and floods. These actions include not only physicians but also other healthcare professionals competent to act in these catastrophic settings by attending to medical emergencies, alleviating suffering, and organizing preventive measures. They are also characterized by their impartiality regarding cultural and political characteristics and, in cases of armed conflict, preference for any faction [33].

War ethics is not a synonym of bioethics, and even less so of medical ethics. Recently, many books and papers have been inquiring into and offering reflections on the multiple problems and dilemmas arising from modern technologies to develop more-powerful and lethal military resources. One primary situation is the justification to kill, surpassing the politics and ideologies that give it a moral and ethical dimension. An interesting precedent is the deep and profound McMahan analysis on the theme, focusing, first at all, on the significance of human identity as living beings and persons and, after that, on the necessary egalitarian considerations and the consequent respect due to every human. He does not analyze armed conflicts or war situations, his interest instead being the ethics of abortion and euthanasia, but he leaves considerations on the ethics of killing and, for our considerations here, on war killing. Also, the differences in the perception of justice and the characteristics that make a war valid need to be assessed through different cultural and political traditions [34].

The first question is whether war can be just and in what sense. Is war just when reconquering or recovering invaded territory? Is a defensive war just? From an ethical perspective, the apparent response is no. From a bioethics point of view, limiting the respect for and the preservation of human life is less permissible. From a utopian bioethical viewpoint, wars do not have a reason to exist. In the real world, nevertheless, dilemmas are related to eliminating it in the measure of the possible or, at least, developing a bioethical consciousness in every person and especially in those who make important political and economically influential decisions. The just war theory is a predominant one, prevalent in wide circles of political thinkers, and it intends to justify a variety of pretexts for making armed aggression against other countries. This is one specific problem. Another, and the most serious, is the unchaining of armed conflicts from war declarations, which attempts to eliminate all considerations of justice, international conventions, and so on.

Thus, in armed conflicts, a serious ethical, now bioethical, dilemma emerges: Who are the people toward whom killing is not only licit but correct? The first capital questions are whether a right to kill exists, whether it supersedes the personal fundamental human right to life, and when and in what conditions killing is permissible. In cases of declared wars, one might logically conclude that killing soldiers of the enemy army is not only correct but desirable [35]. This situation reinforces the conflict discussed previously among the military physicians' bioethical professional obligations toward an injured soldier, friend or enemy, and the possible and most frequently imposed duty to attend only to their army comrades or, at most, prioritize compatriots in every triage decision.

From the implicit necessity to kill, this time concerning specifically combatants, the dilemmas are when to kill and who the victim would be. Some authors have posed the question about which enemies deserve to die, including not only enemy combatants but also civilians who are related to the enemy or suspected of being related to them. In the same work, another author expresses various ways to think about this grave problem. As an example, we have the consideration of defensive rights and killing as a consequence [36]; this takes in account that many times dealing out death is a fact in humanitarian interventions, which presents a moral asymmetry between innocent threats and their victims, where many times killing is derived from defensive acts, which is plausibly a just action of harm [37].

Due to the development of new technologies and weapons designed with them, some unethical killing has been present in recent armed conflicts. Attacks with killer robots and with drones do not make any distinction between combatants and non-combatants, enemies and friends, or civilians and soldiers [38].

However, the field of bioethical reflection remains open to the vast array of problems arising from current global conflicts, with many countries either invading others or being invaded, and most of them have suffered from internal armed struggles of various origins and forms. Could the ethics of war change? Will it be less deadly and more humanistic? The question is open [39], and the eternal pressures of colonialist, ideological, economic, or openly destructive interests have been increasingly building or destroying partnerships without any ethical, global, or medical bioethical concern or remorse [40].

Actual bioethics cannot be left aside, less so in cases of suffering, the destruction of housing, torture, and death affecting civilian populations. According to bioethics, we ought to prioritize attending to civilians' care, both by civilian organizations and by military healthcare givers, independent of their friendship or enmity. Beneficence, justice, and protecting vulnerability are principles that must be followed, in line with the proposals of Garrafa [41] and Kottow [42] on what they call respectively intervention or hard bioethics and protection bioethics, which call for the putting into practice bioethical principles and rules before taking actions that might infringe on the rights and protection of civilian populations.

The reflection that we presented above on killer robots comes here freshly to our attention with the continued development of fighting machines, autonomous to human control. Before they are put into action, the lives of civilians, the total loss of their rights, and their loss of security need to be taken into account and prioritized for the respect, protection, and promotion of human dignity [43].

The severe effects on the environment, social living conditions, and public health from conditions in war and every internal armed conflict need to be taken in account [44]. Here, in consideration of space, we must insist on the importance of bioethics regarding the duties and obligations of all people—military, civilians, politicians—to take into account and prioritize the necessary actions to minimize these deleterious effects on individuals, public health, and the environment.

11.2.3 Nuclear Weapons

From the Hiroshima and Nagasaki atomic bombardment to the nuclear projectiles directed at and exploded on nuclear plants in Iran, even in 2025, the menace of nuclear war has always been present and has increased the possibility of using them. Killing a multitude of noncombatants and innocent people is inexcusable in bioethics, signifying a total rupture of bioethical principles, rules, and virtues. The fight against not only its employment but also research on its production, amelioration, and economically attractive perspectives of commerce is a priority. This is, we insist, nothing new, but it persists and increases day by day [45].

Yet in the twentieth century, the 1970s specifically, one ethical and humanistic association formed, the International Physicians Preventing Nuclear War (IPPNW), headed by Bernard Lown, a Boston cardiologist, and Evghenij Chazov, also a cardiologist, seconded in Mexico and Latino American countries by Manuel Velasco Suárez, a neurosurgeon and the founder of the National Bioethics Commission (Comisión Nacional de Bioética) and the National Mexican of Bioethics Academy (Academia Nacional Mexicana de Bioética) in his country. With more than 1700 physicians affiliated, they were awarded the 1985 Nobel Peace Prize. His work was infused with strong bioethical principles, originated by the considerations of nuclear weapons' immediate lethal consequences on civil populations and the grave health problems that develop in the subsequent years. He aimed to prevent future highly lethal armed conflicts and had a deep preoccupation with the possibility of a third world war [46].

Bioethical reflections and considerations on war ethics need to be open to different ethical theories and points of view and maintain a severe critical position, not being limited to conventional approximations but always looking to develop an awareness and conscientiousness about the immense multiplicity of dilemmas and offering possible and viable ways to reach at least partial solutions.

11.2.3.1 Bioethics: Confronting Genocide

Genocide has been defined as "any act seeking the partial or total destruction of a national, racial, ethnic, political, or religious group with a particular purpose." It includes death, damage to physical or moral integrity, extermination, or the adoption of measures seeking to undermine births within a specific community. Recently, Cardinal Juan Sandoval Íñiguez had included among genocide acts the destruction of embryos. The term *genocide* was coined by Raphael Lemkin in 1944. He proposed this neologism in light of the criminal acts of the Nazi regime against the Jewish people during World War II [47]. In 1948, the Convention for the Prevention of Genocide and Sanctions against Genocide and 50 years later, in 1998, the International Penal Court Statute penalized it [48]. The awful fact is that independently of all these international conventions and theoretical measures, genocide continues to be a frequent crime against humanity. For us, a fundamental question is how, coming from bioethics, we can fortify common sense, positive political decisions, and civilian and social institutions and groups in the fight against these generic injustices.

Twentieth-century history is full of these tragedies: The Turkish Armenian genocide in 1915, precisely documented by Kazantzakis's literary work [49]. One of the most notorious and condemnable ones was that of Nazi Germany against Jews, homosexuals, and any non-Aryan people, in the most reductive definition of Aryan, with the perverse help of physicians and research scientists, sometimes through free will and sometimes forced [50, 51]. More-recent examples include those of Rwanda and Somalia [52, 53], the unconceivable massacres following Yugoslavia's disintegration after the death of Marshal Tito, the Russian bombardments on Ukraine, and the undeclared war on Gaza, among some others. The menace of genocide in a global world, with its immensely crowded cities, is a terrible possibility [54]. These flagrant violations of human rights are deadly menaces to the "humanitarian illusion" [55] are flagrant testimonies of the partial and inadequate conceptions of humans, followed by intentions to annihilate any one not within the most powerful ethnic group, excused by the ignorance—many times manipulated, especially by totalitarian regimes—between fellow humans and strangers [56, 57]. Genocide, in declared or undeclared wars, involves some critical bioethical dilemmas, among which the complete annulment of human rights may be most relevant [58]. Then, it is an intolerable situation from a bioethical perspective, breaking all its principles. Simply, genocide is unacceptable.

Political compromises and responsibilities are mainly responsible for the conditions that lead to genocide. National governments must exercise good conscience and prioritize respecting and promoting human dignity, safeguarding all human beings, and limiting killing, including in wartime. In front of the insensibility of national governments and the partnerships developed among some of them, without a minimum consideration for human lives, is a lamentable fact: the ineffectiveness of international organizations in influencing the wars of the past hundred years and genocides [59].

In this desperate situation, the hope that bioethical consciousness, with the consequent understanding of the value of humanity, would form and develop must be maintained and promoted as a way to contend with this loss of human values due to the priority given to economic factors. The activity of corporations imbued with humanitarian attitudes and professional ethics obligations is a relevant factor, but it always takes form in unbalanced power dynamics. In this context, we must insist on developing any possible actions with medical ethics, legality, and political will in mind, with an openness to criticism, and with honest self-criticism.

11.2.3.2 Armed Conflicts in "Peacetime"

We also want to mention the indubitable growing presence of grave armed conflicts within national territories, sometimes predominant in rural contexts, sometimes in the villages or cities. In official peacetime, violence, killing, and arbitrary maleficent acts against diverse people or groups of people amount to developing warlike conditions. Guerrilla groups, cartels, and increased urban and rural delinquency are common today in many countries and societies. Another grave reality is terrorism, an unjustifiable problem, each day more frequent, which has been seen as a situation justifying torture and even death [60].

We consider these conditions to be real war situations and confront them with the same bioethical considerations employed in war ethics. Self-defense could be a reasonable pretext to kill. But inside what limits is self-defense reasonable? For governments, is imposing military discipline on the total civilian population licit? Who or what justifies it—military dictatorships? And in what terms and inside what limits is such an imposition reasonable?

An interesting study on one particular situation has been conducted on the presence of very heavy tensions in Mexico in the armed conflict of the Pacific resistance in the early 1990s, with the Ejército Zapatista de Liberación Nacional (EZLN) movement. The dilemma here is the possibility of exercising human rights, including those of the rebels and the civilian inhabitants of the rebel-occupied territories, in rebel attacks, army actions, and negotiations [61].

These armed conflicts in peacetime will be the subject of special bioethical reflections and be open to discussion and critical considerations within particular conditions and possibilities, always searching for new options for open and established actions with real bioethical referents.

11.3 Proposals from Bioethics

Following everything briefly discussed in the preceding paragraphs, we consider the promotion of a bioethical culture in all social spheres. Also, insistence on it must emanate from directive groups. Having a shared place is a strong recommendation for bioethics courses in schools, at all levels, but with reinforced commitment in universities, especially in the faculties of health, law, and political science, in addition to philosophy.

The essential perspectives to consider, with the basic consideration and promotion of bioethical principles and virtues, are human rights, their legal promotion, and the protection of vulnerability. The creation of an ambiance and the necessary conditions for the renaissance and growth of an essential humanism, individual and social, based on the recognition, protection, and growth of human dignity is an indispensable bioethical obligation. The resultant acquired responsibility, put into action at all levels—individuals, professional groups, social and governmental organizations, and national and international organizations—is a bioethical imperative more prominent in war and armed conflict than in peacetime.

References

1. Fischer H, Keynes G. The apology and treatise of Ambroise Paré, containing the voyages made into divers places, with many of his writings upon surgery. Gesnerus. 1954;11:49–50.
2. Gross ML. Bioethics and armed conflict: moral dilemmas of medicine and war. Cambridge, MA: MIT Press; 2006.
3. Larrey DJ. (Dominique Jean), baron. Mémoires de chirurgie militaire, et campagnes de D.J. Larrey [Internet]. US: Oxford University; 1812. Available from: https://archive.org/details/mmoiresdechirur00larrgoog/page/4/mode/2up.
4. André Soubiran MG. Le Baron Larrey, Chirurgien de Napoléon. Rev Int Croix-Rouge 2010/04/19 ed. 1967;49:296–7.
5. Viesca CT, Ramos Rocha de Viesca M. La atenzione chirurgica al interno dei campi di battaglia. Antecedenti al lavoro della Croce Rossa, 2017.
6. Bret P. Jean Marchioni, Place à Monsieur Larrey, chirurgien de la garde impériale ; Henri Ducoulombier, Un chirurgien de la Grande Armée. Le baron Pierre-François Percy et Jean Flahaut, Charles-Louis Cadet de Gassicourt (1769–1821), bâ. Ann Hist Révolution Fr. 2009;235–8.
7. Pringle J, Benjamin H, Pringle J, Rush B. Observations on the diseases of the army. Philadelphia: Anthony Finley/Fry and Kammerer, Printers; 1812.
8. Barton C. The Red Cross: a history of this remarkable international movement in the interest of humanity. Allbany: Gryphon Editions; 1898.
9. Sidel VW. Warfare II: medicine and war. Reich Warren Thomas Ed Encycl Bioeth Revis Ed. New York: Simon and Schuster Macmillan; 1995.
10. Pellegrino ED, Thomasma DC. For the patient's good: the restoration of beneficence in health care. New York: Oxford University Press; 1988.
11. Pellegrino ED, Thomasma DC. The virtues in medical practice. New York: Oxford University Press; 1993.
12. Vastyan EA. Warfare: medicine and war. Reich Warren T Ed Encycl Bioeth Vol 4. New York: Free Press; 1978.
13. Rosebury T. Medical ethics and biological warfare. Perspect Biol Med. 1963;6:512–23.
14. Sidel VW. Warfare IV: chemical and biological weapons. Reich Warren Thomas Ed Encycl Bioeth Revis Ed. New York: Simon and Schuster Macmillan; 1995.
15. Kevorkian J. Prescription: medicide: the goodness of planned death. Nachdr. Amherst: Prometheus Books; 1999.
16. Lynch A. Triage and justice: the ethics of rationing life-saving medical resources Gerald R. Winslow. Berkeley: University of California Press; 1982. p. 240. $19.95 (U.S.). Dialogue. 1983;22:754–6.
17. Lazar S, Frowe H, editors. The Oxford handbook of ethics of war. New York: Oxford University Press; 2018.
18. Gonzáles J. Dignidad Humana. In: Tealdi JC, editor. Dicc Latinoam Bioét. UNESCO, Red Latinoamericana y del Caribe de Bioética, Universidad Nacional de Colombia; 2008. p. 277.
19. Beyleveld D, Brownsword R. Human dignity in bioethics and biolaw. Oxford/New York: Oxford University Press; 2001.
20. Sulmasy DP. Dignity and bioethics: history, theory, and selected applications. In: Schulman A, editor. Hum Dign Bioeth Essays Comm Pres Counc Bioeth. President's Council on Bioethics; 2008.
21. Spiegelberg H. Human dignity: a challenge to contemporary philosophy. In: Spiegelberg H, editor. Steppingstones ethics fellow exist essays 1944–1983 [Internet]. Dordrecht: Springer Netherlands; 1986. p. 175–98. https://doi.org/10.1007/978-94-009-4337-7_10.
22. Kass L. Life, liberty and the defense of dignity: the challenge for bioethics. 1. paperback ed. San Francisco: Encounter Books; 2004.
23. Kass L. Defending human dignity. Washington, D.C.: Report No.: Human Dignity and Bioethics: Essays Commissioned by the President's Council on Bioethics; 2008.

24. Churchland P. Human dignity from a neurophilosophical perspective. In: Lanigan BT, editor. Hum Dign Bioeth. New York: Nova Science Publishers, 2008.
25. Nussbaum M. Human dignity and political entitlements. In: Schulman A, editor. Hum Dign Bioeth Essays Comm Pres Counc Bioeth. President's Council on Bioethics; 2008.
26. Lanigan BT, editor. Human dignity and bioethics. New York: Nova Science Publishers, Inc; 2008.
27. Schulman A. Bioethics and the question of human dignity. In: Schulman A, editor. Hum Dign Bioeth Essays Comm Pres Counc Bioeth. President's Council on Bioethics; 2008.
28. Isaac JC. A new guarantee on earth: Hannah Arendt on human dignity and the politics of human rights. Am Polit Sci Rev. 1996;90:61–73.
29. ten Have H. Vulnerability: challenging bioethics. London/New York: Routledge, Taylor & Francis Group; 2016.
30. Allhoff F, editor. Physicians at war: the dual-loyalties challenge. New York: Springer; 2008.
31. Kamm FM. Ethics for enemies: terror, torture, and war. Oxford: Oxford university press; 2011.
32. Abuelaish I. I shall not hate: a Gaza doctor's journey. Toronto: Vintage Canada; 2011.
33. Slim H. Humanitarian ethics: a guide to the morality of aid in war and disaster. New York: Oxford University Press; 2015.
34. Sorabji R, Rodin D, editors. The ethics of war: shared problems in different traditions. Aldershot, Hants/Burlington: Ashgate; 2006.
35. Jenkins R, Robillard M, Strawser BJ, editors. Who should die? The ethics of killing in war. New York: Oxford University Press; 2018.
36. Benbaji Y. Pre-emptive rules and the scope of defensive rights. In: Strawser BJ, Jenkins R, Robillard M, editors. Who should ethics kill war [Internet]. Oxford University Press; 2018 . [cited 2025 Jul 11]. https://doi.org/10.1093/oso/9780190495657.003.0003.
37. Christie L. Distributing death in humanitarian interventions. Who should ethics kill war [Internet]. Oxford University Press; 2018. https://doi.org/10.1093/oso/9780190495657.003.0010.
38. Burri S. What is the moral problem with killer robots? In: Strawser BJ, Jenkins R, Robillard M, editors. Who should ethics kill war [Internet]. Oxford University Press; 2018. [cited 2025 Jul 11]. https://doi.org/10.1093/oso/9780190495657.003.0009.
39. Coker C. Ethics and war in the 21st century. New York: Routledge; 2008.
40. Pfaff CA. Proxy war ethics: the norms of partnering in great power competition [Internet]. Cham: Springer Nature Switzerland; 2024. [cited 2025 Jul 11]. Available from: https://link.springer.com/10.1007/978-3-031-50458-7.
41. Garrafa V, Porto D. Intervention bioethics: a proposal for peripheral countries in a context of power and injustice. Bioethics. 2003;17:399–416.
42. Kottow M. Bioética de protección. In: Tealdi JC, editor. Dicc Latinoam Bioét. UNESCO, Rcd Latinoamericana y del Caribe de Bioética, Universidad Nacional de Colombia; 2008. p. 161–4.
43. Saxon D. Fighting machines: autonomous weapons and human dignity. Philadelphia: University of Pennsylvania Press; 2022.
44. Toole MJ. Warfare III: Publich health and war. Reich Warren Thomas Ed Encycl Bioeth Revis Ed. New York: Simon and Schuster Macmillan; 1995.
45. Forrow L. Warfare: III. Nuclear warfare. Reich Warren Thomas Ed Encycl Bioeth Revis Ed. New York: Simon and Schuster Macmillan; 1995.
46. Migliorini P. I premi Nobel: la vita, le scoperte e i successi dei premiati in fisica, chimica, medicina, letteratura, pace, economia dal 1901 al 2011. 1st ed. Milano: BookTime; 2011.
47. Tealdi JC, editor. Diccionario latinoamericano de bioética. Bogotá: UNESCO [u.a.]; 2008.
48. Colorado GG. Diccionario enciclopedico de bioetica. Editorial Trillas Sa De Cv; 2011.
49. Winter JM, editor. America and the Armenian genocide of 1915. New York: Cambridge University Press; 2003.
50. Bloxham D. Genocide on trial: war crimes trials and the formation of Holocaust history and memory. Oxford: Oxford University Press; 2001.

51. Freyhofer HH. The Nuremberg medical trial: the holocaust and the origin of the Nuremberg Medical Code [Internet]. New York: Peter Lang Verlag; 2005. Available from: https://www.peterlang.com/document/1054410.
52. Kiernan B. Blood and soil: a world history of genocide and extermination from Sparta to Darfur. New Haven/London: Yale University Press; 2009.
53. Guichaoua A. From war to genocide: criminal politics in Rwanda, 1990–1994. Wisconsin: University of Wisconsin Press; 2015.
54. Campbell K. Genocide and the Global Village. 1st ed. 2001. New York: Palgrave Macmillan US; 2001.
55. Debré B. L'illusion humanitaire. Paris: Plon; 1997.
56. Dubiel H, Motzkin G, editors. The lesser evil: moral approaches to genocide practices. London/New York: Routledge; 2004.
57. Shelton D, editor. Encyclopedia of genocide and crimes against humanity. Detroit: Macmillan Reference; 2005.
58. Peterson D, Herman ES, editors. The politics of genocide. New York: Monthly Review Press; 2011.
59. LeBor A. Complicity with Evil [Internet]. Yale University Press; 2006. [cited 2025 Jul 11]. Available from: http://www.jstor.org/stable/j.ctt1npvrk.
60. McPhersonLK.Legalism,justice,andthewaronterrorism[Internet].OxfordUniversityPress;2017 [cited 2025 Jul 11]. Available from: https://academic.oup.com/book/2081/chapter/142004780
61. Arias Marín Alán. Conflicto, resistencia y derechos humanos. 1a ed. México: Comisión Nacional de los Derechos Humanos; 2015.

Part IV
Artificial Intelligence and Neuroethics

Chapter 12
Bioethical Challenges of Regulating Neurorights in the Field of Neuroscience

José Damián Carrillo-Ruiz , Selene Márquez-Chaparro, José Rodrigo Carrillo-Márquez , and María Fernanda Carrillo-Márquez

12.1 Introduction

Human beings are by nature social entities, which is why we interact and relate to each other in various spheres of coexistence, such as family, work, sports, and altruistic bonds. However, not all these human relationships produce legal consequences; not all these relationships create, transform, modify, or extinguish rights and obligations. On the other hand, social processes have evolved, and these changes have shaped new rules of coexistence, some of which are already recognized by law, while others are merely social.

Likewise, advances in technology and health sciences have opened up a range of alternatives for humanity to interact with one another, as well as to improve the life expectations of the human race; therefore, having the possibility of a better quality of life. Considering what was mentioned before, we will begin by examining the legal framework that will lead to the final proposal of this writing to include Civil Law to recognize neurorights. The objective of this chapter is to analyze the ethical

J. D. Carrillo-Ruiz (✉)
Neuroscience Coordination, BINCA Group, Research Direction and Neurosurgery Service, Mexico Anahuac University & Mexico General Hospital, Mexico City, Mexico
e-mail: damian.carrillo@anahuac.mx

S. Márquez-Chaparro
Pontifical John Paul II Institute-Mexico Anahuac University, Mexico City, Mexico
e-mail: selene.marquez@anahuac.mx

J. R. Carrillo-Márquez
Westhill University, Mexico City, Mexico

M. F. Carrillo-Márquez
Faculty of Health Sciences, Mexico Anahuac University, Mexico City, Mexico
e-mail: maria.carrilloma@anahuac.mx

D. Cerdio-Domínguez, M. E. de los Ríos-Uriarte (eds.), *Current Bioethics*, New Paradigms in Healthcare, https://doi.org/10.1007/978-3-032-25294-4_12

challenges that arise when uniting the disciplines of neuroscience and law with their result on subjects, all to regulate neurorights [1].

12.2 What Is Law?

Citing the French jurist Eugène Petit: "Man is gifted with free will, that allows him to develop his natural faculties. But, in Society, this freedom is necessarily limited by respect for the freedom of others" [2]. This is where the notion of "Legal Order" becomes relevant, which is why we will begin by understanding the concept of "Law" before entering the matter of "Neurorights."

The word "Law" is an equivocal term, meaning it has several connotations. For purposes of this document, we will define Law as: "The set of legal norms that regulate human life in society" [3], understood from the personal sphere to the relationship with other human beings and with the State itself. It is an undeniable fact that with the passing of time, human relationships have evolved; thus, the conception of the legal framework and norms of human coexistence needs to change. The history of law studies this evolution, from the legal point of view, as well as the changes that have been necessary to adjust to the social needs of each era. From what we have mentioned, we can deduce that the legal order that governs different regions of the world varies according to beliefs, customs, religions, and multiple human interactions.

This is why there are various legal systems, with the two most recognized worldwide being Roman law and Saxon law. On the one hand: Roman Law, which is the denomination given to the set of laws and norms that allowed the organization of society in the times of Ancient Rome. It is the legal system that governed between 753 B.C. and the 5th century A.D., when the end of the Roman Empire occurred, later retaken by the French and Spanish. Today, this law has permeated and is the legal basis of most Latin American and Latin European societies. On the other hand, Saxon Law, also called "Common Law," refers to a legal system based on decisions made by courts. Judges dictate their decisions based on the principles, customs, and jurisprudence of the country in which it is applied. This law is the legal system in force in England and in most countries that belong to the Anglo-Saxon tradition, such as the United States, Canada, or Australia [4].

Another necessary premise to understand the action of law is that "reality conditions law," according to it processes are created to adjust legal norms to the needs, realities, and advances of social, labor, corporate, and medical relationships, among many others; these are forged over time in each different place, adjusting the legal norm to the facts or social advances that arise in each country or region. Just as the legal norm will be adjusted and updated in matters of general and special medical advances, primarily in matters of neurorights.

12.3 The Person

The person is currently defined in many ways, depending on whether it is approached by philosophy, law, or bioethics. Since old times, it was of utmost importance to know who the person was and what the consequences of their behavior with themselves or their lack of respect toward others were. The Greeks endowed actors with masks characterized as different characters. It is at this moment that the word person was born; the masks used in those classical theaters were called "*prosopon*". Some years later, in the Roman Empire, the term "*personare*" was created, which means "to sound"; the same concept that made an impact to the Law, until nowadays [5].

The classical definition of what a person is can be attributed to Boethius, who defines the person as "individual substance of rational nature" [6]. In outlining this definition, the following is recognized. The personal being will always be substance, meaning it exists and can sustain itself [7]. The person is an individual being since it is a unique and indivisible being; no human being is equal to another because only they are who they have to be. Human nature is biological, pointing out that man is obligated to follow the natural laws that constitute him. In classical philosophy, the concept of "*physis*" or nature is understood as the concept of "*telos*" or finality [8]. Finally, rationality differentiates man from any being that is reduced to its corporeality; the person is a being that is intelligible and conscious of its existence [9]. In this way, man cannot only be understood from one of his dimensions; the person is constituted by a body and by a soul (in its classical concept of anima); a biological part and a spiritual part. This means that the person is a unique being who uses their intelligence to self-determine. Following this argument, personality reflects what the being is in its substance, communicating through its circumstances what relates to its ontology.

One of the simplest and most enriching definitions of person is the one philosophy provides with Emmanuel Mounier, who describes the person as an "incarnate spirit." It is a being that is composed of an ethereal part and a body [10], so there exists a body (represented in the case of Neuroethics essentially by the brain) and its spirit, defined by Max Scheler as the center of values. On the one hand, there is the corporeal part, which helps us express ourselves with others. Without a body, one being could not be human since it would be an ethereal entelechy; our humanity necessarily requires a body and a spirit. Emmanuel Mounier complements Boethius's definition by emphasizing that the human person is composed of body and spirit; it is understood as a different being, physically by accidents and spiritually by that immaterial essence [10]. The spirit inclines them to choose good and moral, same it incites man to seek virtue and truth; to recognize their neighbor and themselves always seeking human dignity. From the spirit comes freedom since, if there is volitional determination, the body expresses what it is commanded to do. Due to this existential characteristic, man desires to surpass himself to live a fuller life by fulfilling his aspirations [11]. As a synthesis, the person is a merge between organic (brain), mental (mind) and spiritual (spirit) (Figure).

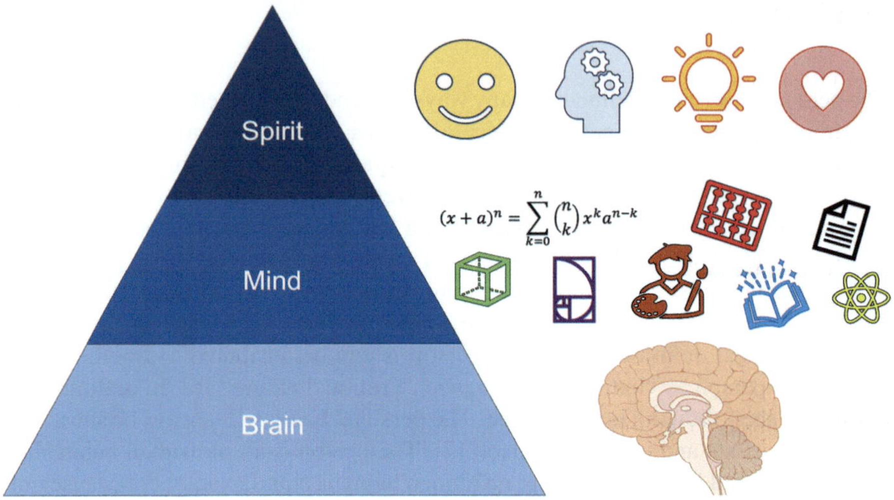

12.4 Legal Concept of Person

As mentioned, in philosophy, the "human" person is defined as a being gifted with reason, conscience, and the capability for moral choices. It is a being that reflects on its existence, seeks meaning and values, confronting fundamental questions about its purpose and destiny. The concept "*Person*" in its most general definition refers to the human being, legally cited as "*Natural Person*"; however, from the legal perspective, this definition also extends to other entities created by man to associate to perform lawful legal acts, whether commercial, social, or charitable; which integrate various partners to that common and lawful purpose, such as commercial companies, civil societies including the State itself [12].

For Law, the person is the subject of rights and obligations constituted as prerogatives, whether a natural person or a legal person; both must comply with the attributes of personality to be able to interact in the legal sphere. Their attributes are: Name, Nationality, Patrimony, Civil status (only in the case of natural persons), Legal capacity, and Residence.

12.5 Be and Ought to Be

In the context of law, "being" refers to the reality of human conduct, while "ought to be" implies the obligations imposed by legal norms. "Being law" relates to the content of the law, while "ought to be law" refers to justice and equity in its application. There are laws of being, which are natural, and laws of ought to be, which are social. The distinction between both concepts is fundamental to understand ethics and morality in the field of Law.

Aristotle said that the human being is a political animal and cannot live excluded from society, since being human implies a social nature. Therefore, the society where the human being must develop can be defined as "a conventional institution that groups a set of people who are governed by mandates that we call norms." The coexistence that occurs within a society is susceptible to being modified at will, and it is governed by conventionalisms; a group of people agrees on what norms they must comply with. As we can observe from the two spheres in which the human being develops, there are laws: the first laws that occur within nature we will call laws of being, and those that occur within the social sphere we will call laws of ought to be.

12.6 Rights and Obligations

Law in the broad sense is the set of legal norms that impose rights and obligations, establishing the bases of social coexistence, and whose objective is to provide all members of society with the minimums of security, certainty, equality, freedom, and justice. To perform these functions, these two concepts exist. An obligation is a demand or commitment to which a person is subjected for some cause, and which makes them act in a predetermined manner, either undertaking conduct or abstaining from it. This can be of moral, or legal. For example, "tax obligations" are called the taxes and payments which citizens commit to contribute to the State and the maintenance of the public sector. In contrast, a right is a power or faculty that the law recognizes in individuals, through which people can demand or do something. Thus, there are political rights, such as voting, and civil rights where freedom or property is involved. Most importantly, they are protected by the State through laws.

12.7 Neuroscience

Neuroscience is the discipline that lead to deep knowledge of the nervous system, brain processes, spinal cord, and nerves; which include Neuropsychology, Neurophysiology, Neurobiochemistry, Neurology, Neurosurgery, and Neuropsychiatry. However, recent discoveries have greatly expanded this spectrum, including Neuroimaging, Neuroimmunology, Neuroethology, Neurogenetics, or Neuroproteomics; but even surprisingly, there are new branches that have first been in formation and then as a dynamic process of development such as Neurophilosophy and Neuroethics [13].

Medicine advances day by day with more surprising discoveries and goals, with spectacular technological development being an evident fact in our time [14]. Perhaps one of these parts of science that has advanced the most has been neuroscience, and this is undoubtedly due to the development of scientific achievements, as never before seen in humanity. In 1990, U.S. President George H. Bush, in his speech to the U.S. Congress on the "Decade of the Brain Project," stated the following: "The human brain is a 3-pound mass containing nerve cells that control our activity, being one of the most magnificent and mysterious wonders of creation. The seat of our human intelligence, the interpreter of our senses, and the controller of movements, which is an incredible organ that continues to intrigue scientists and like-minded men" [15]. Thus, as John Horgan would say in 1999, in his book about the mind about the brain: "A typical brain has 100 trillion cells, almost the same number as the stars in the Milky Way. Each of the cells is connected to another 100,000 more. The synapses between cells are converted into hormones and neurotransmitters that modulate transmission signals, and synapses constantly form and dissolve, strengthen and weaken in response to new experiences" [16].

The importance of neuroscience is indisputable; a great number of Nobel Prizes have been awarded to researchers in this area. Chronologically, we can mention below those names, with their respective discoveries, some important figures such as: Santiago Ramón y Cajal (1952) [17] and Camilo Golgi (1989) [18], with the description of the structure of the Nervous System and the Neuronal Theory, in 1906. Sir Charles Sherrington (1913) [19] and Edgar Adrian (1936) [20], with the determination of the functions of neurons through nerves in 1932. Sir Henry Dale (1927) [21] and Otto Loewi (1957) [22], with how the chemical transmission of nerve impulses in the brain and spinal cord develops, in 1936. Joseph Erlanger [23] and Herbert S. Gasser (1958) [24], with the highly functional impact of nerve fibers, in 1944. Walter Rudolf Hess (1973) [25], for the discovery of the brain in the activities of internal organs, and Egas Moniz (1954) [26], with leucotomy (Psychosurgery) in psychoses, in 1949. Sir John Eccles [27], with Alan L. Hodgkin and Andrew F. Huxley (1953) [28], who exemplified the mechanisms involved in the excitation and inhibition of the Central and Peripheral Nervous System, in 1963. Sir Bernard Katz (1939) [29], Ulf Von Euler (1952) [30], and Jules Axelrod (1972) [31], with the definition of the concept of neurotransmission and the mechanism of storage and inactivation in synaptic transmission, in 1970. Roger Guillemin (1978) [32] and Andrew Schally [33] identified the production of peptide hormones in the brain in 1977. For 1979, the prize was awarded to Allan Cormack (1980) and Geoffrey N. Hounsfield (1973) [34], for the discovery of computed tomography, and for the first time, the brain is observed without the need to open the skull. Subsequently, Dr. Roger Sperry [35] studied the functional differentiation of brain hemispheres in 1981, along with David Hubel and Torsten Wiesel (1960) [36], on how visual information is processed. In 1986, Rita Levi-Montalcini was awarded this honor for the discovery of nerve growth factor, which is the substance that makes neurons grow and develop [37]. In 1991, Erwin Neher and Bert Sakmann (1992) [38] performed the identification of ionic channels and their isolated electrical currents in neurons. Subsequently, Stanley Prusiner made the discovery of prions in neurodegenerative diseases in 1997 [39]. For the year 2000, Drs. Arvid Carlsson (1993) [40], Paul Greengard [41], and Eric Kandel [42] studied signal transduction in the nervous system on dopamine and discovered a neuronal mechanism of memory. The prize was awarded in 2003, for the discovery of magnetic resonance to Drs. Paul C. Lauterbur (1989) [43] and Sir Peter Mansfield [44], through which the details of brain anatomy are observed through Magnetic Resonance Imaging. In 2004, the contribution was for the discovery of receptors for smell and the organization of the olfactory system, given to Richard Axel (2005) [45] and Linda B. Buck (1993) [46]. In 2014, the winners were John O'Keefe [47], May-Britt Moser and Edvard Moser [48] for their discovery of the spatial position of organisms detected in the brain. Recently, in 2021, the laureates corresponded to Ardem Patapoutian and David Julius for the discovery of receptors for chronic pain.

Brain exploration in recent years has spread exponentially. After the year 2000, the number of contributions obtained has not stopped; this is why we call "Decade of the Mind." From the years 2000 to 2010, we completed all these discoveries made in the previous decade. However, there are discrepancies in this regard between this organicist brain point and the mental part among the scientific community and philosophers. Schwartz JF & Begley S., in their book "The mind and brain," from 2003, stated the following: "It has been named the Decade of the Brain… having this name is better than the Decade of the Mind. That is why it is more in the brain than in the mind that scientists and like-minded people solve questions, test theories of our gray matter to know the roots of personality and temperament, mental illness and mood, sexual identity, and even a predilection for fine food" [49]. The famous physicist and science popularizer Carl Sagan would express it in his book "The Dragons of Eden": "My fundamental idea about the brain is that it works—in what we sometimes call 'mind'—because of its anatomy and physiology, and nothing more" [50]. But on the other hand, Nobel Prize winner Sir John Eccles, along with philosopher Daniel Robinson, in their book "The Wonder of Being Human," in 1984, expressed it differently: "We observe promissory materialism as a superstition without rational foundation. The more we discover about the brain, the clearer we distinguish between brain events and mental phenomena, and the more wonderful it makes the brain and mental phenomena realize. Promissory materialism is simply a religious belief held by materialistic dogmatism that frequently confuses its religion with its science" [51]. In this same line, Mario Beauregard, Canadian neuroscientist and reporter Denyse O'Leary, in their book "The Spiritual Brain," in 2007, express the following: "The brain is not the mind, it is the accessible organ to connect the mind to the rest of the universe. By analogy, Olympic swimming events require a athlete in the pool, but the pool does not create the events, it simply makes them possible" [52] and Ramón Lucas Lucas in his book: "Anthropology and Bioethical Problems" establishes the following: "What is the relationship between intelligence and brain? It is not a causal relationship but an instrumental condition. That is, intelligence uses the brain to think but does not think with the brain. To be able to think, intelligence needs the brain and the senses to provide the material to think. Without this material, intelligence cannot operate; therefore, intelligence when thinking uses the brain as a condition" [53]. Again, citing Beauregard & O'Leary, who point out that "the discipline of neuroscience today is materialist," the mind is simply the physical work of the brain. Materialists think that the distinction between the mind as an immaterial entity and the brain as an organ of the body has no real basis. The mind is assumed to be only an illusion generated by the work of neurons, and even for them, the term mind does not mean that it exists [52]. Francis Crick, also Nobel Prize winner in 1962, expressed in his book published in 1997, "The Astonishing Hypothesis," that: "Our highly developed brain after all, did not evolve under the pressure of discovering scientific truths, but only to allow us to be intelligent enough to survive and leave descendants" [54], and in this regard, Harvard neuropsychologist Steve Pinker expressed: "The idea that our thoughts, sensations, joys and failures consist totally and exclusively of activity in brain tissue. Consciousness does not reside in an eternal soul that uses the brain as a personal digital assistant; consciousness is brain activity" [55]. In this way, there are conflicting opinions continuing with the dualism that was initially commented on between the mind and brain. There are people in favor of concrete thinking about the brain, and there are others who support a second vision focused on the ethereal aspects of the mind.

During the past century, research provided elements that made the brain and higher mental functions able to be understood by psychologists and neuroscientists [56]. Based on this, at this moment, we seek to "understand how psychotherapeutic interventions work in the same way as we try to understand the mechanism of psychopharmacological interventions," it is proposed that "psychoanalysis has a model of the mind that can be useful for this" [57] and in turn "psychoanalysis can be revitalized through a closer relationship… with cognitive neurosciences…" [58].

The dualistic conception of the human being is a central theme for the philosophy of mind. Cartesian dualism proposes the existence of the mind and body constituted by two substances of distinct natures. Currently, it is proposed that the mind is a product of brain activity and that mental disorders are a product of its dysfunction at the level of specific neural circuits [59].

With the "Decade of the Brain" project (1990–2000), Neurosciences had greater presence and demonstrated their potential utility for studying the origin of mental illnesses and the mechanisms of action of therapeutic agents used for their treatment. As already mentioned, the "Decade of the Brain Project" was an initiative sponsored by the Library of Congress (LC) and the National Institute of Mental Health (NIMH) of the United States. Its central objective was the strengthening of a general awareness about the benefits of the scientific study of the brain [60, 61].

Neuroimaging studies have made evident the effects of psychotherapy, integrating them into current theories of Psychology. To what degree do genetics influence a person's behavior and thinking, and how much does the environment? This question was central to Freud, since he gave capital importance to childhood experiences, while accepting the existence of a constitutional predisposition to diseases. The joint research of two scientific disciplines, on the one hand, Psychoanalysis with its epigenetic proposal and Neurosciences with Genetics, have allowed to elucidate to some degree an answer [62].

12.8 Ethics and Bioethics

Ethics is the discipline that emanates from logic and establishes in human conduct the application of the concepts of good and evil, being a universal concept. Bioethics is the part of ethics that relates to natural sciences focused on health, and it is in this area that it becomes involved in making correct or incorrect decisions about the actions of health professionals in patient illness. The first time this term was coined was by Fritz Jahr, who gave it an ecological vision, in his treatise written on "Bioethics: a look at the ethical relationship of men with animals and plants." Later, in the 1970s, Van Potter would retake this same concept, creating a bridge between scientific and humanistic knowledge. The encyclopedia of Bioethics defines it as: "The science that has its relationship with moral values and principles applied to health sciences and biology." Currently, bioethics has taken a dominant role in hospitals in different committees, since, by law, in countries, bioethical hospital committees and research ethics committees have been established. All based on the Declaration of Human Rights of 1948, by the United Nations, and the Helsinki Declaration of 1950. In addition to the different conferences and codes in which each country has been adjusting with international treaties in its constitution and its health codes: this has made medical praxis be fed back not only by its laws, norms, and regulations, but by concrete recommendations in dilemmas (problems in which the partial or total physical integrity of the patient is exposed and whose resolution implies the use of principles and values), in research protocols and daily clinical life in the care of health personnel concerning their patients. Bioethics is the crossroads between medicine, law, and philosophy, mainly where the proposals are to resolve such dilemmas in the doctor–patient relationship, whether in the office, in the operating room, in the laboratory, in the rehabilitation room, or some other clinical or hospital space for the benefit of the patient. There may also be a contradiction due to the divergence of values between health personnel and the patient, so it employs a respectful and tolerant dialogue.

12.9 Neuroethics

For the first time, the term was used in 1989 by a neurologist named Ronald E. Cranford, the neurologist physician who functioned as a clinical/surgical professional and a neuroethicist. He provided his points of view in ethics committees. Subsequently, a group of philosophers and doctors proposed different concepts in this regard. Adina Roskies, Judy Illes, and Martha Farah stand out in 2002, each establishing their definition separately; together, it would be an emerging discipline of ethics in conjunction with neuroscience. On the other hand, Jonathan Moreno, in 2003, and Thomas Fuchs and Walter Gannon, in 2006, would also establish concepts about neuroethics, its repercussion once again of ethics in neuroscience and in society [63]. In a very clear definition, Adina Roskies establishes the following: "It is the part of Ethics dedicated to the Nervous System, to studies on neuroscience and the uses of neurotechnologies." Currently, this field has been expanding gradually, involving all the aspects that have already been explained in neuroscience, with a focus on technology and on the diagnosis and treatment of patients [64].

Strictly speaking, neuroethics would have to do only with neuroscience and the applications used by it and generate technologies and knowledge. Neurobioethics would oversee the relationship between discoveries and treatments in human beings for clinical and therapeutic purposes, but sometimes this line can be nuanced, and both terms can be used interchangeably.

12.10 Neurolaw

The evolution of neuroscience and neurotechnology has been increasing and developing. This phenomenon will not stop; just as the world advances in discoveries under water and outside the earth's atmosphere, so neuroscientists (doctors, biologists, biomedical engineers, programmers, among others) are discovering new techniques and procedures that will help patients improve ailments that are now difficult to diagnose and treat [65].

In a matter of a few decades, electronic and cybernetic devices will be developed, capable of decoding the information from our brain and introducing themselves into the depths of our mind. The use of technology capable of such power is a double-edged sword because, while it will enable great advances in fields like criminology by discovering if a suspect is the author of a crime, it will also raise serious ethical questions like being able to modify our memories or manipulate our decision-making [66].

In the evolution of social relations, medical and technological advances, the so-called "Neuro Rights" or brain rights have already emerged, which can be defined as: "A new international legal framework of human rights, specifically intended to protect the brain and its activity as advances in neurotechnology occur" [67]. It is a concept that can be understood in two aspects: The first is mental privacy, for example, people's brain data are treated with confidentiality compared to an organ transplant. The second is the right to identity, maintaining the individuality of people [68].

They are also comprehended as the right to cognitive freedom, mental privacy, mental integrity, psychological continuity (or identity self-perception), equitable access to brain enhancement, protection against biases, autonomy of will, and personal self-determination. These are new contents and legal protections, recognizable, enforceable, and unbreachable barriers need to be regulated against new risks and threats from technological developments and their application in human beings [69].

The advances and developments in science and technology for a society like ours open the door to a whole world of solutions for problems that now lack resolution. This is the case of neurotechnology, a discipline that combines neuroscience with new scientific discoveries and that generates the possibility of revolutionizing our understanding and interaction with the human brain, opening a world of possibilities.

12.11 Challenges of Neurolaw

In this panorama, neuroscience develops knowledge that can answer very specific questions in different fields of the nervous system using cutting-edge technology. So, the question the researcher asks is: Can I do it? and try to supply the deficiencies in diagnosis, treatment, and prognosis of the patient. Then comes the ethical part, which questions whether it was previously achieved or reached, and the adequate thing to ask is: Should I do it? knowing in advance whether to continue with said clinical research with the patient, mainly for their benefit. Finally, making a chronological sequence would come with a regulatory part where the question is: Should I regulate this conduct and convert it into a norm or law? And this is for the benefit of the person and society in the last case. You cannot always simultaneously have a scientific advance, alongside the ethical and legal, and sometimes, in a very staggered way, the legal part, to try to control neuroscientific discoveries that could even harm individuals. That is why there is a need to talk about neuroethics and neurolaw. Neurolaw arises from neuroscience and is immediately its predecessor, which would be neuroethics.

12.12 Development of Neurorights

Recently, an academic corpus has been constructed from the scientific knowledge developed in the course of recent years, mainly from the already mentioned decade of the brain from 1990 to 2000, and the decade of the mind, from 2000 to 2010, in addition to the following 15 years in which the breadth of technological and scientific baggage about the brain and mind has allowed knowing in greater depth some aspects of how the acting of individuals works, mainly regarding their free will and their autonomy, aspects that are also studied by law and parallel by ethics. So, neurosciences study the structures of the central and peripheral nervous system, as well as the mind. From these elements, neuroethics emerges as a response to questions in the field of neuroscience and aspects of human behavior, its study being very new. This has been amplifying, and it has come to also include law in its relationship with said acts of individuals; but involving neuroscientific aspects in which the brain and mind interact with their circuits to understand what these rights and obligations are, from the organic point of view and their influence in concrete cases in the legal field.

Neurotechnology is currently being developed alongside this understanding of the human brain and artificial intelligences, as well as the consideration of possibilities until now unthinkable that generate an important focus of attention; from neurofeedback devices that help us improve our well-being to brain-computer interfaces that allow people with functional diversity to communicate and control electronic devices; neurotechnology will have a fundamental role in improving our quality of life [70].

The evolution of neuroscience and neurotechnology is a reality that will not stop. Scientists and engineers in the field of medicine are discovering new techniques and procedures that will help doctors and patients improve ailments that are now difficult to diagnose. It is a matter of a few decades for electronic devices capable of decoding the information from our brain and introducing themselves into the depths of our mind to be invented. Medically, an enormous achievement; however, in opposition in the legal field, it would affect free will, as well as in the internal forum of consent of a subject to create legal consequences [71].

The use of technology capable of such power is a double-edged sword since, while it will suppose great advances in fields like criminology, discovering that a suspect is the author of a crime, it also raises serious ethical questions like being able to modify our memories or manipulate our decision-making.

Neurorights are the new international legal framework in the field of human rights and refer to the legal and ethical protection of safeguarding the brain and its activity, the mind, and privacy of thought, as a consequence of the challenges neurotechnology and neuroscience bring.

These rights are based on interdisciplinarity between different scientific branches (biology, psychology, neurology, among others) with the legal system, since neurotechnology could be used to decipher and manipulate mental processes. A clear example is the Kernel Flow device, which allows "mapping" brain activity through magnetic resonance scanners. Here is where law becomes involved, since with legal intervention, it seeks to protect the privacy of the human mind and mental autonomy [72, 74].

12.13 International Recognition of Neurorights

The concept of neurorights has been developed by the NeuroRights Initiative organization, which focuses on developing a code of ethics that governs scientists involved in neurotechnology, such as those that record information from brain activity or interfere with it. The Neurorights Foundation corporation has agreed on what the five neurorights are: (1) Right to preservation of personal identity, (2) Right to non-interference in freedom of decision, (3) Right to privacy of neuronal data, (4) Right to equity in brain capacity improvement, and (5) Right to protection against discriminatory biases derived from algorithms (Neurorights Foundation).

First, the era in which we live is known as the digital era, in which the way we learn, interact with other people, and work is based on technology, the relationship between technological advances, philosophy, and the duty to create ethical codes. In other words, the regulations through which scientists are governed to carry out technologies that potentially can change society under an approach that guarantees and protects human rights.

An example of research projects similar to the "BRAIN" Initiative, promoted by President Obama in 2013, to promote public research for the development of neurotechnology and Artificial Intelligence is Facebook's "Brain to Text" project that is building a noninvasive brain–computer interface to decode human thoughts at a speed of 100 words per minute and write them on the computer screen.

12.14 Mexico's Proposals

Mexico has joined the international trend of legislating in favor of these rights, showing concern for the lack of regulatory frameworks that protect user privacy against neurotechnologies. As a result, it has stayed informed about international legislation in favor of neurorights.

In 2023, Congresswoman María Eugenia Hernández Pérez participated in the "Forum on Neurorights, Advances and Challenges in Latin America," organized by the Chilean foundation Kamanau, where she highlighted advances in defending and creating a regulatory framework that protects neurorights in Mexico.

Additionally, on January 25th, 2024, the National Council of Humanities, Sciences, and Technology (Conahcyt) and the Chamber of Deputies of the Congress of the Union held a meeting called "Neurorights in the face of the mechanization of the world," in which the legal, ethical, and social implications and impact of these rights in Mexico were analyzed.

Likewise, the Director of the National Center for Human Rights (CENADEH), Master Rosy Laura Castellanos Mariano, emphasized: "Technology and advances must respond to daily life, but humanely, recovering what society needs from a humanistic perspective. Thus, defending rights means finding the limits and scope that these rights themselves can have."

In 2025, a group made up of Dr. Karen Herrera Ferrá from the Mexican Society of Neuroethics, together with then-Congresswoman Hernández, presented the main points of a scientific/governmental committee that would support the implementation and development of neurorights in Mexico. This shows the importance of discussing this topic in governments and its possible international repercussions in different countries around the world [73].

12.15 Conclusions

Neurorights emerge from the understanding of law on the one hand and neuroscience on the other. However, in its depth, Law is built upon the fundamental concepts of the person, being, and ought-to-be, and normativity encompassing rights and obligations. Neurosciences, as the knowledge of the central and peripheral nervous systems together with the mind, have found allies in ethics and bioethics to project neuroethics and integrate with technology to establish new paradigms in the evolution of our society, which is reflected in the emergence of neurorights.

References

1. Wszalek J. Ethical and legal concerns associated with the comprehension of legal language and concepts. AJOB Neurosci. 2017;8(1):26–36.
2. Petit E. Tratado elemental de derecho romano. 23rd ed. México: Ed Porrúa; 2007.
3. Villoro TM. Estudio del derecho. México: Ed Porrúa; 1987.
4. Stein PG. Roman law, common law, and civil law. Tulane Law Rev. 1991;66:1591–605.
5. Rico ÁF. De las personas. México: Ed. Porrúa Porrúa; 2018.
6. Boecio S. In: España FC, editor. Sobre la persona y las dos naturalezas. BAC; 1979. p. 1343.
7. Aristóteles. Metafísica. México: Ed Porrúa; 2004.
8. Aristóteles. Ética a Nicómaco. México: Ed Alianza; 2014.
9. Aristóteles. Tratados de lógica. Mexico: Ed Porrúa; 2019.
10. Mounier E. Esprit. España: Ed. Caparros; 1997.
11. Carrillo-Ruiz J, Carrillo-Márquez J, Jiménez-Ponce F, García-Muñoz L, Navarro-Olvera J, Beltrán J. Neurobioética personalista: su consideración en la psicocirugía. Neurocir Neurochirurgie Démo Flanc. 2020;29:117–35.
12. Código Civil Federal de México. Artículo 25. Ciudad de México: Cámara de Diputados del H. Congreso de la Unión; 2024.
13. Roskies AL. Neuroethics for the new millennium. Neuron. 2002;35(1):21–3.
14. Roldan J. Ética médica. México: Ed Trillas; 1990.
15. Rhoton AL Jr. Neurosurgery in the decade of the brain. J Neurosurg. 1990;73(4):487–95.
16. Horgan J. The undiscovered mind: how the human brain defies replication, medication and explanation. New York: Free Press Ed; 1999.
17. Ramón y Cajal S. Structure and connections of neurons. Bull Los Angel Neuro Soc. 1952;17(1–2):5–46.
18. Golgi C. On the structure of nerve cells. J Microsc. 1989;155(1):3–7.
19. Sherrington C. Further observations on the production of reflex stepping by combination of reflex excitation with reflex inhibition. J Physiol. 1913;47(3):196–214.
20. Adrian E. The nervous system. Science. 1936;84(2178):274–8.
21. Dale HH, Richards AN. The depressor (vasodilator) action of adrenaline. J Physiol. 1927;63(3):201–10.
22. Loewi O. On the background of the discovery of neurochemical transmission. J Mt Sinai Hosp N Y. 1957;24(6):1014–6.
23. Erlanger J, Schoepfle GM. A study of nerve degeneration and regeneration. Am J Physiol. 1946;147(3):550–81.
24. Gasser HS. The postspike positivity of unmedullated fibers of dorsal root origin. J Gen Physiol. 1958;41(4):613–32.
25. Hess WR. Principles of cerebral organization. Experientia. 1973;19:511–4.
26. Moniz E. I succeeded in performing the prefrontal leukotomy. J Clin Exp Psychopathol. 1954;15(4):373–9.
27. Eccles JC, Magladery JW. The excitation and response of smooth muscle. J Physiol. 1937;90(1):31–67.
28. Hodgkin AL, Huxley AF. Movement of radioactive potassium and membrane current in a giant axon. J Physiol. 1953;121(2):403–14.
29. Katz B. Nerve excitation by high-frequency alternating current. J Physiol. 1939;96(2):202–24.
30. Von Euler U. Some aspects of the clinical physiology of noradrenaline. Scand J Clin Lab Invest. 1952;4(4):254–62.
31. Axelrod J. Catecholamines. N Engl J Med. 1972;287(5):237–42.
32. Guillemin R. Biochemical and physiological correlates of hypothalamic peptides. Res Publ Assoc Res Nerv Ment Dis. 1978;56:155–94.
33. Schally AV, Bowers CY, Locke W. Neurohumoral functions of the hypothalamus. Am J Med Sci. 1964;248:79–101.

34. Di Chiro G, Brooks RA. The 1979 Nobel prize in physiology or medicine. Science. 1979;206(4422):1060–2. https://doi.org/10.1126/science.386516.
35. Sperry C, Gadsden CP, Rodriguez C, Bach LM. Miniature subcutaneous frequency-modulated transmitter for brain potentials. Science. 1961;134(3488):1423–4.
36. Hubel DH, Wiesel TN. Receptive fields of optic nerve fibres in the spider monkey. J Physiol. 1960;154:572–80.
37. Levi-Montalcini R. The nerve growth factor. Ann N Y Acad Sci. 1964;118:149–70.
38. Neher E, Sakmann B. The patch clamp technique. Sci Am. 1992;266(3):44–51.
39. Prusiner SB. Prions: novel infectious pathogens. Adv Virus Res. 1984;29:1–56.
40. Carlsson A. On the neuronal circuitries and neurotransmitters involved in the control of locomotor activity. J Neural Transm Suppl. 1993;40:1–12.
41. Greengard P, Kuo JF, Miyamoto E. Studies on the mechanism of action of cyclic AMP in nervous and other tissues. Adv Enzyme Regul. 1970;9:113–25.
42. Kandel ER, Frazier WT, Watchtel H. Organization of inhibition in abdominal ganglion of Aplysia. J Neurophysiol. 1969;32(4):496–508.
43. Lauterbur PC. Image formation by induced local interactions. Clin Orthop Relat Res. 1989;244:3–6.
44. Mansfield P, Maudsley AA. Medical imaging by NMR. Br J Radiol. 1977;50(591):188–94.
45. Axel R. Scents and sensibility: a molecular logic of olfactory perception (Nobel lecture). Angew Chem Int Ed Engl. 2005;44(38):6110–27.
46. Buck LB. Receptor diversity and spatial patterning in the mammalian olfactory system. Ciba Found Symp. 1993;179:51–64.
47. O'Keefe J, Burgess N, Donnett JG, Jeffery KJ, Maguire EA. Place cells, navigational accuracy, and the human hippocampus. Philos Trans R Soc Lond B Biol Sci. 1998;353(1373):1333–40.
48. Moser EI, Roudi Y, Witter MP, Kentros C, Bonhoeffer T, Moser MB. Grid cells and cortical representation. Nat Rev Neurosci. 2014;15(7):466–81.
49. Schwartz JM, Begley S. The mind and brain: neuroplasticity and the power of mental force. New York: Harper Collins; 2003.
50. Sagan C. The dragons of Eden: speculations on the nature of human intelligence. New York: Random House; 1977.
51. Eccles J, Robinson D. The wonder of being human: our brain and our mind. New York: Free Press; 1984.
52. Beauregard M, O'Leary D. The spiritual brain. Toronto: Harper Collins Publishers; 2007.
53. Lucas Lucas R. Antropología y problemas bioéticos. Madrid: BAC; 2008.
54. Crick F. The astonishing hypothesis: the scientific search for the soul. New York: Oxford University Press; 1989.
55. Pinker S. The mystery of consciousness. Time. 2007;169(5):58–70.
56. Olds DD. Psychoanalysis and the neurosciences. In: Gabbard GO, Litowitz BE, Williams P, editors. Textbook of psychoanalysis. 2nd ed. Arlington: American Psychiatric Publishing; 2012.
57. Kandel ER. Psychotherapy and the single synapse: the impact of psychiatric thought on neurobiological research. N Engl J Med. 1979;301:1028–37.
58. Kandel ER. Biology and the future of psychoanalysis: a new intellectual framework for psychiatry revisited. Am J Psychiatry. 1999;156:505–24.
59. Andreasen NC. Linking mind and brain in the study of mental illnesses: a project for a scientific psychopathology. Science. 1997;275:1586–93.
60. Human Brain Project. Human Brain Project website. 2014. Available from: https://www.humanbrainproject.eu/. Accessed 30 June 2025.
61. Deza L. La década del cerebro. Rev Neuropsiquiatr. 1990;53:77–8.
62. Caspi A, Moffitt T. Gene-environment interactions in psychiatry: joining forces with neuroscience. Nat Rev Neurosci. 2006;7:583–90.
63. Fuchs T. Neurobiology and psychotherapy: an emerging dialogue. Curr Opin Psychiatry. 2004;17:479–85.

64. Ruiz S, Valera L, Ramos P, Sitaram R. Neurorights in the constitution: from neurotechnology to ethics and politics. Philos Trans R Soc Lond B Biol Sci. 1915;2024(379):20230098.
65. Vidal C. Neurotechnologies under the eye of bioethics. eNeuro. 2022;9(3):ENEURO.0072-22.2022.
66. Chandler JA. Inferring mental states from brain data: ethico-legal questions about social uses of brain data. Hastings Cent Rep. 2025;55(1):22–32.
67. de Asís R. Sobre la propuesta de los neuroderechos. Derechos Libert Rev Filosof Derechos Hum. 2022;47:51–70. https://doi.org/10.20318/dyl.2022.6873.
68. Szoszkiewicz Ł, Yuste R. Mental privacy: navigating risks, rights and regulation. EMBO Rep. 2025;26(14):3469–73.
69. Wajnerman PA. Is your neural data part of your mind? Exploring the conceptual basis of mental privacy. Minds Mach. 2022;32(2):395–415.
70. Meynen G, Van de Pol N, Tesink V, Ligthart S. Neurotechnology to reduce recidivism: ethical and legal challenges. Handb Clin Neurol. 2023;197:265–76.
71. Bublitz JC. What an international declaration on neurotechnologies and human rights could look like: ideas, suggestions, desiderata. AJOB Neurosci. 2024;15(2):96–112.
72. Istace T. Neurorights: the debate about new legal safeguards to protect the mind. Issues Law Med. 2022;37(1):95–114.
73. Herrera-Ferrá K, Muñoz JM, Becerril A, García-López E, Marinaro JÁ, Sánchez Hernández LR, et al. The regulation of neurotechnology: the neurorights bill in Mexico. Lancet Psychiatry. 2025;12(2):88–90. Erratum in: Lancet Psychiatry. 2024;11(12):e14
74. Carrillo-Ruiz JD, Carrillo-Márquez JR, Beltrán JQ, Jiménez-Ponce F, García-Muñoz L, Navarro-Olvera JL, Márquez-Franco R, Velasco F. Innovative perspectives in limbic surgery using deep brain stimulation. Front Neurosci. 2023;17:1167244. doi: 10.3389/fnins.2023.1167244.

Chapter 13
Bioethics and Artificial Intelligence

María Elizabeth de los Ríos-Uriarte and José Sols Lucia

13.1 Introduction

In thesis number nine of his "Theses on the Concept of History," [1] the German philosopher Walter Benjamin describes the fatality of progress as a hurricane-force wind that impedes the beating of the wings of the angel of history and forces it to look back on the past with an air of pessimism, finding ruin upon ruin. This pessimistic sense of progress and its consequent critique, which accompanies the philosophers of the Frankfurt School, is still shared today in the face of the accelerated advance of science and technology, which, although beneficial, appears lurking and fearsome in other respects. Therefore, from the perspective of applied ethics and bioethics, it is necessary to critically analyze some of its manifestations, such as the rapid and irreversible incorporation of Artificial Intelligence (AI) into our lives.

This study is a continuation of our book *Bioethics of Artificial Intelligence*, [2] in which we present ethical reflections in general and bioethical reflections in particular on the applications of AI in the field of health. Firstly, we will present a justification for addressing the development of AI in the field of health from the perspective of bioethics and its principles. Secondly, we will explore some of its applications in medicine and its possible future projections. And thirdly, we will analyze its risks and benefits, as well as other possible consequences of its application.

M. E. de los Ríos-Uriarte (✉)
School of Law, Universidad Autónoma de Querétaro, Querétaro, México
e-mail: maria.elizabeth.delosrios@uaq.mx

J. S. Lucia
Universidad Iberoamericana, Ciudad de México, México
e-mail: jose.sols@ibero.mx

D. Cerdio-Domínguez, M. E. de los Ríos-Uriarte (eds.), *Current Bioethics*, New Paradigms in Healthcare, https://doi.org/10.1007/978-3-032-25294-4_13

13.2 Justification of the Bioethical Approach

From its integration into preventive, curative, and rehabilitative medicine to applications accessible on mobile phones to measure vital signs, alert to risky medical conditions, and communicate with family members in case of medical emergencies, the benefits of AI are manifold and promise to be even more so. However, we also glimpse risks and anthropological and ethical deviations. For this reason, it is essential to resort to both inductive and deductive methodologies for its analysis.

Bioethics is a science with a long history. In 1927, the notion appeared from the mouth of a German Protestant theologian, Fritz Jahr, [3] as a legitimate concern for the behavior of human beings in the face of realities such as scientific advancement through animal experimentation. In the 1970s, the American doctor of Dutch-origin Van Ransselaer Potter [4] promoted the term "bioethics" in relation to caring for the environment and medical action in problematic cases.

Similarly, some international patient rights movements, such as those arising from paradigmatic cases such as that of Nancy Cruzan [5] or that of Karen Anne Quinlan [6], forced us to think about a *set of* principles and considerations that would safeguard the integral health of individuals, especially terminal patients, as well as to promote and strengthen their capacities and warn about possible malpractices that could harm them.

In this way, bioethics is defined as "the systematic study of human behavior in the field of health and life sciences in the light of moral and ethical principles and values" [7]. In 1995, this definition was corrected: "Bioethics is the systematic study of the moral dimensions—including moral vision, decisions, behavior, guidelines, etc.—of the life and health sciences, using a variety of ethical methodologies in an interdisciplinary approach" [8]. Its main task is to gather the empirical data necessary to define the ethical values at stake and to define courses of action that protect the values of life, health, dignity, or bodily integrity of the person. In this way, one of the areas that has been most enriched by the incorporation of AI-driven systems and technologies is medicine, and for this reason, we consider it necessary to illuminate reflection based on bioethical principles.

Advances in AI are relentless. Our analysis is based on a time frame that always leaves the door open to new considerations. From its original intentions,[1] bioethics aims to build a bridge between the present and the future, based on ethics that prevent outrages and abuses of human dignity.

[1] This is the title of Potter's most famous work. Cf. Potter, Van Ransselaer, *Bioethics: A Bridge to the Future* (New Jersey: Prentice Hall, 1971).

13.3 Applications of AI in the Field of Health and Its Future Projections

One of the challenges on the horizon for both medicine and bioethics is the "Four Ps Medicine [9] (*precision*, *personalization*, *prediction*, and *prevention*), which attempts to anticipate the timely diagnosis of diseases or conditions that, based on personalized studies, can be treated with schemes adapted to each person's particular genetics as well as to their social, environmental, and family conditions, providing them with the greatest possible benefit for the restoration of their health or the maintenance of their quality of life and assuming that not all patients are the same.

AI is a huge help in the prevention and care of diseases [10]. The first applications of AI for timely diagnosis or preventive control studies were in the area of clinical pathology, in clinical blood studies in laboratories, as well as in radiology and other imaging studies, providing the opportunity to obtain parameters, calculations, patterns, and associations more quickly and with greater precision from data obtained in the laboratory, all of which constitutes a greater probability of success in medical action [11]. In preventive medicine, we already have electronic devices for measuring vital signs with their consequent recording and wireless transmission to the corresponding doctor, as well as the activation of alert systems to a network of family and friends in case of high-risk medical conditions, falls, or states of unconsciousness.

However, there is a fine line between benefit and risk. For example, the development of robots that care for the elderly can raise serious questions about the replacement of the doctor–patient relationship [12]. Robots are increasingly entering the field of medicine, to the point of becoming nurses or surgeons. There are robots that "care" for and "talk" to the elderly. They seem to show empathy and have feelings toward them, for example, by listening to their stories over and over again and asking them to retell them.

There are contact lenses that monitor glaucoma and automatically apply medication to the eye. There are insulin pumps attached to the body that apply the doses the person needs when alterations in glucose level are detected.

The problems begin when we move from "therapeutic medicine[2] to "desire medicine" [13], the aim of which is not to *relieve pain* or *cure,* but to *improve* something that is already good. This is the case with human *enhancement* interventions,[3] which originated in England and later in the United States in the 1960s. We are referring

---

[2] This is defined as the ethical legitimacy of intervening in a part of a person's body when it is sick and puts their health or life at risk. When the part to be intervened does not warrant it, the action is not ethically justified. See Elio Sgreccia, *Personalist Bioethics*, Philadelphia: National Catholic Bioethics Center, 2012, p. 180.

[3] *Enhancement* is a series of medical actions and interventions (invasive or not) that aim to improve capacities or abilities to exceed human standards. Cf. Savulescu, Julian, "Justice, Fairness, and Enhancement," in Murray, Thomas H. and Chuan, Voo Teck, (2020). *The Ethics of Sports Technologies and Human Enhancement.* Oxfordshire: Routledge, 2020. https://doi.org/10.4324/9781003075004.

here to *transhumanism*,[4] a philosophical, cultural, and artistic movement that promotes the improvement of the human species through the introduction of new technologies into our bodies. Those in favor of this type of improvement assume that our condition is imperfect and that, therefore, we must continue along the evolutionary line through the progressive incorporation of technology into the human body. Transhumanists[5] promote this idea of unlimited improvement of the human body, and posthumanists speak directly of a new species,[6] the posthumans. Both want to go beyond the limits of the human condition due to the rejection of essences and ontological fixism [14]. They have a triple objective: superintelligence [15], super well-being [16] and super longevity [17].

Superintelligence The aim is to improve intellectual and cognitive abilities through the intake of drugs such as methylphenidate, [18] which allows for higher levels of concentration and memory, or through specific intervention in certain parts of the brain to stimulate its activity with electrodes or other devices. Similarly, the brain–computer interface [19] (specific sensors placed in the brain) allows the brain to remotely control electronic prostheses. This has been a great help for people with reduced or no mobility and has allowed them to carry out activities such as walking or eating independently.

Super Wellness Drugs are being used to make people feel happiness, pleasure, and well-being; [20] there are applications that offer the possibility of experiencing pleasant or comforting scenarios, such as virtual reality or the metaverse [21].

Super Longevity The consumption of certain drugs lengthens DNA telomeres or the design of certain diets and routines in such a way as to delay the effects of aging on the skin, DNA, and the brain.[7] Will we be able to live to be 500 years old? Will we never be able to die? Since ancient times, the desire for immortality has been present in the human spirit, and even today, it survives with blind faith in science and technology.

[4] Cf. Postigo Solana, Elena, "Transhumanism, Human Enhancement, and the Bioethical Challenges of Emerging Technologies for the 21st Century": *Cuadernos de Bioética* 32/105 (2021), pp. 133–139. https://doi.org/10.30444/CB.92.

[5] Transhumans are human beings with biotechnological bodily parts, or with interventions to enhance their natural physical, psychological, or neurological capacities, who still retain their human essence and, therefore, their corporeal limits. Cf. Gómez Álvarez, José Enrique, "Post and Transhumans": *Medicina y Ética* 34/2, pp. 521–530.

[6] Despite the ambiguity of the term even today, we can define the *posthuman* as a hypothetical human being who has undergone technological intervention in such a way that their physical and corporeal limits have been completely surpassed, to the point that they appear to be a member of a new species. Cf. Rugo, Daniele, "Posthuman," in *Oxford Research Encyclopedia of Literature,* oxfordre.com

[7] We suggest reviewing the "Don't Die" Project, in which aging has been reversed through the design of a specific lifestyle called "Blueprint." See dontdie.bryanjohnson.com

What is the purpose of medicine? According to the Hippocratic Oath, [22] that purpose is the cure and relief of the conditions and ailments that afflict people and put their lives or physical integrity at risk. Any other purpose would be contrary to good medical practice. However, according to transhumanism, the purpose of medicine is to improve the human species indefinitely.

AI applied to the field of health sciences provides very useful tools in terms of speed in processing large amounts of data, efficiency, and accuracy, as well as in relation to timely diagnoses and treatments, and therefore contributes to personalized and highly accurate medicine that offers a longer life expectancy even in remote or difficult-to-reach places where doctors could not easily reach. Its benefits are many and will continue to be so as long as the doctor–patient relationship is not replaced by machines or devices. Here we argue that interpersonal contact must remain the backbone of medical care.

13.4 Ethical Considerations of the Use of AI in Healthcare

Let us now look at some of the risks of introducing the development of AI into the health sciences, in order to critically analyze its viability for the fulfillment of the ethical principles that truly safeguard and protect the life and health of individuals and communities. Although there are many more, here we will present only four bioethical issues.

13.4.1 Algorithmic Biases

We have already said that the incorporation of AI in healthcare occurred mainly in clinical laboratory studies, as it allowed for greater speed and accuracy in information processing. However, AI does not have autonomy, that is to say, the information it works with comes from human beings who enter it into the system and design it to incorporate more data and execute tasks in a simpler and faster way, but the main source of information is still people [23]. They have biases, which are reflected and increased in the system. These biases have consequences when the results are produced, as they can be based on the information that was given to them and also on what was omitted. For example, false positives occur in some breast cancer diagnoses because a mass is considered abnormal when, due to its characteristics, it simulates a cancerous pattern, when in reality it is not [24]. Bias also occurs in the data entered into a system that refers only to certain groups of people based on their age, skin color, ethnicity, social class, occupation, etc., which the AI associates with adjectives such as "desirable" or "undesirable," in such a way that it encompasses personal characteristics and recognizes the labels with which it has been fed, being able to discriminate against or marginalize those who do not conform to these patterns.

This has been increasingly refined, and therefore, possible biases have been mitigated and reduced through various samples of heterogeneous groups, but it is still a problem that has significant emotional and psychological repercussions [25].

13.4.2 Lack of Spontaneity and Human Creativity

We have talked about the replacement of the doctor–patient relationship by a robot. There are robot surgeons who, although they allow high levels of accuracy and precision, lack properly human faculties such as creativity and spontaneity, which can be decisive in the operating theatre. To claim that the robot is programmed to do what the surgeon controls from outside only that demonstrates that the robot does not have the capacity to act in the face of an unexpected situation that can occur in any surgery, such as uncontrollable bleeding, a previously undetected problem, an unknown anomaly, an irregular heart rhythm, or even an unexpected heart attack. A surgeon does have the intelligence and imagination to resort to resources or protocols that help them to deal with unexpected problems.

Doctors have, or should have, values [26] such as honesty, assertive and efficient communication, empathy, a commitment to improving health in general, and trust. A robot cannot have these, even if it appears to have them. Narrative is important to sustain this link and is made up of stories. Robots have no history. They only act according to a computer program.

In this sense, we believe that it is right for robots to support human medical work, but not to supplant it.

13.4.3 Data Protection

AI systems operate with large amounts of data (*big data*), which in the field of health are sensitive because they are biometric. To date, it has not been possible to protect these data. Cyberspace is porous, and they get hold of it one way or another [27]. This becomes especially sensitive not only because of the susceptibility to identity theft but also because of those applications that are remotely controlled in large centers where these data are stored. Think of scenarios where remote medication administration devices are controlled, such as insulin pumps for diabetic patients. These storage centers can be hacked and the data and algorithms altered, thus causing an increase or a deliberate shortage in the administration of medication that can put at risk the lives of those who use and depend on the devices to maintain their stability. Think of North Korean hackers endangering the lives of thousands of Americans or Europeans by computer manipulation of the technology they carry in their bodies.

In a society in which social utility is prioritized over the ontological value of individuals, many individuals would be discarded because of a "regrettable computer error" in the devices on which they depend, which could end their lives. This leads us to consider a fourth problem, which is the attribution of responsibility.

13.4.4 Attribution of Responsibility

When a mistake is made in medicine, not only can its causes be traced, but it can also be attributed in most cases because there is a doctor or other health professional who caused it and on whom the responsibility and, if necessary, the relevant sanctions fall. But this does not happen when the agents involved in the action are not people, but machines or AI systems that lack freedom and, therefore, rationality.

If the failure was caused by these developments, the question of who would be responsible can have large legal loopholes that leave the door open to irresponsibility and, therefore, to many risks in which people are completely unprotected. For this reason, current regulations [28] propose a delimitation of the actions of each agent in the process of design, development, and use of AI systems to guarantee that, in all their facets, these systems conform to ethical principles, with each party assuming their respective responsibility for the final consequences.

In addition, European Parliament regulations provide for safe spaces for training systems before they are released for use and commercialization in Europe. The aim is to detect possible failures in their execution but also the phases in which they occur in order to identify the people behind each one and to determine and demand stricter regulations that reduce the margin of error.

This does not rule out the possibility that, even if they have been created and designed with specific regulations and training has been provided in these safe spaces, the multiple variables involved in their operation may get out of control and generate some kind of failure, but it does allow for the documentation of all phases and the tracking of their implementation to determine responsibility.

Needless to say, this responsibility will always fall on those who can make real decisions according to their anthropological structure, that is, the rational dimension that, among other things, estimates the risks, benefits, and consequences of the action to be taken and chooses based on that estimation. This is only possible, then, in human beings and, therefore, responsibility can never be attributed to a machine, but always to a person. This once again confirms the anthropological idea underlying any development of AI, which is that it is and always will be human beings who feed AI to generate certain specific results or responses. In other words, machines do not make decisions by themselves; it is human beings who guide and feed them to do so [29].

Thus, the *results are* nothing more than a reflection of the actions, prejudices, decisions, understandings, patterns of association, identification, and rationalization of reality of the people behind each machine that operates with AI.

13.4.5 Inequalities in the Distribution of Benefits and in the Inaccessibility of the Resources and Advances Generated

A final ethical problem that we present in this chapter is the fact that the advances and positive aspects that AI can offer are not distributed equitably to all the people who can benefit enormously from them. This is due, on the one hand, to the fact that most of the resources invested in these developments come from private initiative and, on the other hand, to the fact that they are offered at very high costs.

Regarding the first aspect, investment by private initiative calls into question incentive programs for science and research and the establishment of national AI strategies offered by governments,[8] which, although it has been increasing in recent years, is still behind private investment, [30] which is committed to return on investment, to make their production processes more efficient, and to position themselves in the global market, leaving social welfare and the search for equality in accessibility in the background.

This means that, when AI-powered devices or products are launched and proven effective, their costs rise above people's purchasing power. If we add to this the convergence and intersectoral factors of people's vulnerabilities (e.g., poverty, disability, gender, dialect, isolated rural or urban location, and remoteness), the benefits become practically impossible to obtain.

Fighting for greater public sector budget allocations to invest in AI developments would generate a greater production of these, which would allow a greater supply to satisfy the great demand, and therefore, the costs could be reduced and be accessible to more people.

13.5 Conclusions

The multiple applications of AI in medicine promise significant advances, both in the prevention and prediction of diseases and in their care and monitoring. However, these advances bring with them ethical challenges that demand attention. Bioethics becomes a key tool to ensure that the incorporation of these technologies respects fundamental principles such as human dignity, privacy, justice, and integrity in medical care.

Although AI offers clear benefits, such as faster and more accurate diagnoses, it also poses significant risks, such as algorithmic biases, lack of spontaneity and human creativity in the doctor–patient relationship, or violation of sensitive data

[8] Cf. Guevara-Gómez, A. (2023). NATIONAL ARTIFICIAL INTELLIGENCE STRATEGIES: AN APPROACH FROM A GENDER PERSPECTIVE. *Revista De Gestión Pública,* 12(2), 129–154. Retrieved from https://rhv.uv.cl/index.php/rgp/article/view/4299.

protection. These challenges highlight the need for clear regulations that define responsibilities and promote the ethical use of AI.

In the field of healthcare, the integration of AI should be seen as a *complement to* optimizing medical care, not as a *substitute for* healthcare professionals. Human contact and the essential values of medicine, such as empathy and trust, remain fundamental pillars. Therefore, the challenge ahead of us is to build an interdisciplinary dialogue that allows for a balanced and ethical use of AI, always prioritizing the well-being and rights of individuals.

References

1. Benjamin W. Concepts of the philosophy of history. Buenos Aires: Terramar; 2007.
2. Sols Lucia J, de los Ríos Uriarte ME. Bioethics of artificial intelligence. Madrid: San Pablo–Comillas Pontifical University; 2024.
3. Márquez-Vargas F. Toward a foundation of environmental bioethics from the perspective of Fritz Jahr, Aldo Leopold, and Van Rensselaer Potter. Colomb J Bioeth. 2022;15(2) https://doi.org/10.18270/rcb.v15i2.3009.
4. Potter VR. Bioethics, the science of survival. Perspect Biol Med. 1970;14(1):127–53.
5. Annas GJ. The long dying of Nancy Cruzan. Law Med Health Care. 1991;19(1–2):52–9. https://doi.org/10.1111/j.1748-720X.1991.tb01793.x.
6. Stevens MLT. The Quinlan case revisited: a history of the cultural politics of medicine and the law. J Health Polit Policy Law. 1996;21(2):347–66. https://doi.org/10.1215/03616878-21-2-347.
7. Reich WT. Encyclopedia of bioethics. New York: MacMillan; 1978.
8. Reich WT. Encyclopedia of bioethics. New York: MacMillan; 1995.
9. Hood L, Flores M. A personal view on systems medicine and the emergence of proactive P4 medicine: predictive, preventive, personalized, and participatory. New Biotechnol. 2012;29(6) https://doi.org/10.1016/j.nbt.2012.03.004.
10. Sols J, de los Ríos ME. Bioethics of artificial intelligence. Madrid: San Pablo–Comillas Pontifical University; 2024. p. 37.
11. Ekmekci PE, Arda B. Artificial intelligence and bioethics. New York: Springer; 2020.
12. Sols J, de los Ríos ME. Bioethics of artificial intelligence. Madrid: San Pablo–Comillas Pontifical University; 2024. p. 101.
13. Blasco-Fontecilla H. Medicalization, wish-fulfilling medicine, and disease mongering: toward a brave new world? Rev Clin Esp (Engl Ed). 2014;214(2):104–7.
14. Ravasi G. A panoramic and symbolic look at transhumanism. Pensamiento. 2022;78(298):461–70. https://doi.org/10.14422/pen.v78.i298.y2022.008.
15. Zapata-Muriel FA, Montoya-Zapata S, Montoya-Zapata D. Ethical dilemmas posed by the rise of artificial intelligence: a perspective from transhumanism. Rev Cient Arbitr Multidiscip PENTACIENCIAS. 2023;5(2):94–104. Available from: https://editorialalema.org
16. Chaparro-Gosálbez N. Transhumanism: the limits of the human. La Torre del Virrey. J Cult Stud. 2024;35:1–26.
17. Contreras-Calderón J. Bioethical analysis of the transhumanist movement and the eugenics initiative for an axiological reflection. Apuntes Bioética. 2024;7(1):20–37. https://doi.org/10.35383/apuntes.v7i1.1057.
18. Pinto-Venegas JP. What is neuroenhancement? Towards an enactive understanding of the phenomenon. Rev AJAYU. 2023;21(2):148–55. https://doi.org/10.35319/ajayu.212215.
19. Casas-Martínez ML. Enhancement and transhumanism, an accepted change? Dilemata. 2023;41:17–26.

20. Quevedo-Martín D. The institutionalization of transhumanist thought: proposals and critiques. Polít Soc. 2023;60(3):1–14. https://doi.org/10.5209/poso.80821.
21. Alcívar-Cedeño AK, Bastidas-Logroño DJ, Toctaguano-Cruz SJ, Mora-Marcillo AB. Human-computer interaction in the educational metaverse. Rev Cient Arbitr Multidiscip PENTACIENCIAS. 2023;5(2):94–104. Available from: https://editorialalema.org
22. Mesa-Trujillo D, Espinosa-Ferro Y, García-Mesa I. Reflections on medical bioethics. Rev Méd Electron. 2022;44(2):413–24. Available from: https://scielo.sld.cu
23. García-Vigil JL. Reflections on ethics, human intelligence, and artificial intelligence. Gac Med Mex. 2021;157(3):311–4. https://doi.org/10.24875/gmm.20000818.
24. Amaya-Santos S, Jiménez-Pernett J, Bermúdez-Tamayo C. Health for whom? Intersectionality and biases in artificial intelligence for clinical diagnosis. An Sist Sanit Navar. 2024;47(2) https://doi.org/10.23938/assn.1077.
25. Iturmendi-Rubia JM. Algorithmic discrimination and its impact on human dignity and human rights: special reference to immigrants. Deusto J Hum Rights. 2023;12:257–84. https://doi.org/10.18543/djhr.2910.
26. Campos-Olazábal JP. Medical virtues: a look at some clinical cases. Cuad Bioét. 2022;33(107):111–8.
27. Basil NN, Ambe S, Ekhator C, Fonkem E. Health records database and inherent security concerns: a review of the literature. Cureus. 2022;14(10):e30168. https://doi.org/10.7759/cureus.30168.
28. European Union Artificial Intelligence Act. Available from: https://artificialintelligenceact.eu
29. Esteban-Basáez JM, Mora J. Health and artificial intelligence: how have we evolved? Rev Méd Clín Las Condes. 2022;33(6) https://doi.org/10.1016/j.rmclc.2022.11.003.
30. Jiménez A. Artificial intelligence in health: a local and regional perspective. Medicina. 2022;43(4):620–30. https://doi.org/10.56050/01205498.1650.

Part V
Bioethics Education

Chapter 14
Integral Formation in Bioethics and Research Ethics

Raúl Jiménez Piña and Aida Cristina Coss Rojas

14.1 Introduction

The training of human resources, in any area of knowledge, will always benefit from the acquisition of skills that enable critical and ethical practice, regardless of the profession. This chapter will focus on training provided in the medical field and any other discipline that prioritizes healthcare, since the role of ethics and all scientific and technical knowledge have an inextricable correlation.

It is not possible to separate the teaching of bioethics from the ethical dimension of the educational act itself. Teaching necessarily implies a moral obligation: the teacher acts as a symbolic reference, a role model, and a generator of a dialogic space where students confront their beliefs, values, and emotions.

Despite various approaches suggesting that students reach undergraduate level with a consolidated moral foundation, the experience in clinical settings confront them with dilemmas that go beyond their prior ethical references. In these cases, students often resort to two main sources: the normative documents they are familiar with and the observation of their professional role models. If a process of critical reflection does not accompany the influence of these sources, the risk of reproducing institutionalized vices or unethical behaviors is increased.

R. J. Piña (✉)
Center for Bioethical Knowledge, National Bioethics Commission of Mexico, Ciudad de México, México

A. C. C. Rojas
Academic Development Deputy, National Bioethics Commission of Mexico, Ciudad de México, México

D. Cerdio-Domínguez, M. E. de los Ríos-Uriarte (eds.), *Current Bioethics*, New Paradigms in Healthcare, https://doi.org/10.1007/978-3-032-25294-4_14

For this reason, bioethics, by its deliberative nature, requires a different pedagogical approach than the exact sciences. It is not about learning "correct answers," but rather about practicing ethical thinking, rational argumentation, and informed decision-making in the face of uncertainty [1].

Since the beginning of the twentieth century, our society is recognizable by a dizzying and, seemingly, unstoppable advance of science and technology, which has had an impact on the life on earth and all its dimensions, as well as on social dynamics. This, also, has resulted in a gap between science and its application, and its legal and ethical regulation, which is why it is considered of the utmost relevance that undergraduate students understand that training and learning ethics are a fundamental component in their professional development, since their ethical skills will be tested in their professional daily activities. To overcome these situations, they must have, on the one hand, the skill to identify and analyze the occurrence and appearance of ethical dilemmas in their daily tasks and, on the other hand, know the methodological tools that allow finding possible solutions, within a framework of scientific evidence, rational argumentation, and respect for human rights.

14.2 Historical Foundations of Bioethics and Research Ethics

The concept of bioethics dates back to 1927, with Fritz Jahr's article "Bio-Ethics: A Review of the Ethical Relationships of Humans to Animals and Plants." Drawing from the Kantian imperative, Jahr argued that every living being should be regarded as an end in itself, not merely as a means. This ethical consideration extends to the environment, which deserves moral attention in terms of its use and care [2].

More than 40 years later, in the United States, Van Rensselaer Potter used the concept bioethics in his 1970 article "Bioethics, a Bridge to the Future" to propose an approach focused on the ethical use of biomedical knowledge, combined with the protection of the environment to ensure the future survival of all living beings on Earth [3].

In parallel, the work of André Hellegers in 1971 led to the establishment of the first institute devoted to the study of bioethics—the Kennedy Institute—which marked the beginning of the institutionalization of the discipline. Initially, this institutionalization was primarily oriented toward biomedicine, with a focus on education, clinical care, and research.

Beyond the academic development of these disciplines, the social and political contexts of the time also contributed to their evolution. Paradigmatic cases such as the Nuremberg and Tuskegee experiments, as well as individual situations like those of Karen Ann Quinlan and Terri Schiavo, along with the advocacy for patients' rights and the rapid advances in biomedical technology, made it necessary to create ethical guidelines and standards—such as the Belmont Report and the Declaration of Helsinki, to name just a couple of examples in the field of research. These developments also led to the establishment of clinical ethics committees and the creation of analytical methodologies to address the dilemmas arising in clinical practice.

As a field of study, during the 1970s and 1980s, bioethics became institutionally consolidated in North America and Europe through the creation of hospital ethics committees, interdisciplinary academic centers, and specialized training programs. Early influential publications, such as *Principles of biomedical ethics* by Beauchamp and Childress [4], who proposed a principles framework—autonomy, beneficence, non-maleficence, and justice—became a global reference for ethical analysis.

During the 1990s and 2000s, bioethics experienced both geographical and thematic expansion, incorporating emerging global concerns such as equity in access to healthcare, environmental ethics, the rights of Indigenous peoples, and the impact of biotechnology and genetics. In this context, UNESCO played a key role in promoting the development of international normative instruments, most notably the *Universal Declaration on Bioethics and Human Rights* (2005), which put forward a pluralistic, intercultural, and human rights–based approach with global relevance.

Today, bioethics faces contemporary challenges such as artificial intelligence in healthcare, personalized medicine, genetic editing, global health inequalities, the climate crisis, and the ethical management of pandemics. In this new landscape, the global approach to bioethics requires a critical, interdisciplinary, and contextual perspective—one capable of integrating cultural diversity, social justice, and sustainability into ethical analysis.

The global evolution of bioethics reflects, in sum, a dynamic process of institutionalization, expansion, and transformation, continually shaped by the historical, scientific, geographic, and cultural conditions of contemporary life.

The boundaries of bioethics continue to expand, and the imperative to equip future generations of professionals from diverse disciplines with analytical tools and a critical mindset is a pressing contemporary demand—particularly as social challenges evolve and grow increasingly complex.

14.3 Ethics in the Education of Health Professionals

Currently, the vast majority of schools offering programs related to healthcare have incorporated courses on bioethics and research ethics under various titles. In many cases, these courses are mandatory, although there are still institutions where they are considered elective or of secondary importance. Research on the teaching of these subjects tends to focus on structural aspects—such as their presence in the academic curriculum, timing, course load, or content—without sufficiently addressing the transversal and longitudinal nature that bioethics should adopt as a formative discipline. Thus, a comprehensive approach is not limited to the transmission of theories, models, or normative frameworks but should be oriented toward creating reflective spaces that encourage personal and ethical inquiry by students.

The teaching of bioethics and research ethics cannot be understood as a mechanical extension of pedagogical strategies applied in other fields of knowledge. Its complexity lies in the fact that it challenges deep dimensions of subjectivity, individual morality, and social sensitivity. Although bioethics and research ethics present conceptual differences, the term ethics will be used here to encompass both solely for expository purposes and to streamline reading, limiting its scope to the field of health professions, while maintaining clarity regarding the distinction in their origin, development, and reach, and considering the context of clinical education.

14.4 Ethics of Teaching and Teaching of Ethics

The teaching of ethics requires a fundamental coherence between the content being conveyed and the conduct of the educator. Ethics cannot be effectively taught without embodying ethical teaching practices. In this sense, the act of teaching itself constitutes a moral endeavor—one that demands respect, integrity, and honesty.

Some educators tend to believe that ethical values are the exclusive responsibility of the family environment and that university students should already come "formed" in that realm. While it is true that axiological foundations are often originated in the domestic sphere, the reality of clinical practice—including the educational setting—raises conflicts and dilemmas that demand new responses, not always contained within previously acquired precepts [5].

Students typically resort to two main sources when facing these dilemmas: accessible and observable normative frameworks and the behavioral modeling they receive from their teachers and colleagues. Without a reflective process that allows them to question and critically analyze these references, there is a risk of perpetuating inappropriate practices—even those contrary to fundamental ethical principles. The example set by the educator—through their consistency, attitudes, and decisions—constitutes one of the most powerful mechanisms for the moral formation of students [6].

A common mistake among educators is to assume that their own convictions are ethically unquestionable and evaluate students based on their conformity to those beliefs. Ethics, however, is not reducible to uncritical adherence to a set of rules; rather, it is grounded in critical and independent thinking, rational deliberation, and individual responsibility [5, 6]. Consequently, ethical education should be aimed more at fostering moral autonomy than at imposing dogmas.

14.5 Ethical Training

As mentioned, the internalization of ethics as part of the educational process goes beyond scientific and technical knowledge; therefore, its teaching must also aim at ethical awareness, critical thinking, and an understanding of human motivations and attitudes. It likewise requires personal development in knowledge, skills, and ethical reasoning that are valuable both within and beyond the professional context, as learning these disciplines fosters an openness to other perspectives and values, thereby improving interpersonal coexistence [7, 8].

Bioethics and research ethics, although distinct in their areas of analysis, share a common foundation of skills, attitudes, and knowledge for their application—built upon philosophy but enriched by other disciplines such as medicine, law, psychology, anthropology, and biology. In other words, they are interdisciplinary fields [9]. Therefore, it is logical that teaching and training in these areas should be approached comprehensively, combining not only knowledge from various disciplines but also strategies that enable, first, the identification of ethical dilemmas and problems as they arise, and later, their resolution through well-structured, evidence-based arguments [9, 10].

During the learning and teaching of bioethics and research ethics, it is equally important to address broad and complex issues while also incorporating context-specific problems. This approach helps avoid the fragmentation of knowledge from individual disciplines and fosters an understanding of the complex relationships between various factors that can give rise to ethical dilemmas. Therefore, it is essential to develop skills such as critical and creative thinking, ethical reasoning, decision-making, and effective communication, among others [6, 11].

This relates to the fact that every decision—and consequently, every action—is based on factors that may arise from three main areas, as explained by Sánchez [12]: cognition, approach, and application. The first involves understanding the relevant facts of the situation—that is, the contexts, theories, and so on. The second refers to the assessment of the values involved and their hierarchy. Finally, the third involves proposing an optimal course of action based on the previous factors and in accordance with current laws. Failing to consider any of these areas would lead to ethical reductionism and, therefore, to an incomplete analysis of the situation and its ethical implications.

From this perspective, it is understandable why interdisciplinarity is important for bioethical deliberations, in which different approaches from various fields equally contribute to completing the necessary information from each area. For example, healthcare professionals understand the scientific and physiological facts and consequences of a case, a lawyer would provide the legal perspective, and an anthropologist the cultural context. Therefore, these perspectives should also be included in academic programs [12].

Meanwhile, Osbeck et al. proposed a model of the phases of ethical reasoning that delves into the personal skills required at each stage to carry out the decision-making process as objectively as possible. Therefore, these skills should be practiced throughout bioethics and research ethics education programs. They are summarized in Table 14.1 [13].

As can be seen, teaching and training in bioethics go beyond theoretical knowledge—in science, medicine, philosophy, anthropology, etc.— with this discipline being a clear example of the need for convergence between the humanities and the sciences to understand the practical consequences of the rapid development of technological innovations and their application [14].

Table 14.1 Phases of ethical reasoning

Phase	Knowledge and skills required
Sensitivity	Recognition of potential dilemmas Listening to others' arguments Understanding of the contexts involved Empathy
Judgment	Reflecting on the available information Critical analysis of information Ability to clearly and coherently explain the reasons behind resolutions
Motivation	Analyzing personal values and principles Determining the influence of these values and principles, as well as other subjective motives, in the reasoning process
Implementation	Effective and clear communication of the reasoning behind the chosen resolution within a specific context and the evidence on which it was based

Modified from Osbeck et al. [13]

14.6 Ethics and Learning

It has been noted that students in fields related to healthcare, especially in the early years of undergraduate studies, tend to show limited sensitivity to the ethical aspects of clinical practice, as they rarely must make significant decisions on their own. At this stage, decisions are mainly made by their instructors or superiors, placing students in a passive role. However, when they are presented with ethical problems appropriate to their level of responsibility, they can begin to develop awareness of their own values and start to take ownership of the consequences of their decisions.

Some common dilemmas that can prompt this type of reflection include disclosing their student status to patients; handling a patient's refusal to be treated by a student; accessing clinical information; performing procedures for which they do not feel fully prepared; obeying orders that conflict with their personal values; or witnessing dishonest behavior among peers. The response to these issues cannot be singular or prescribed; rather, it requires personal deliberation informed by general ethical principles, but also by the individual's own value structure.

An additional obstacle is the frequent undervaluation of bioethics as a "soft science" compared to the so-called "hard" biomedical sciences, which calls for a cultural shift within medical education.

14.7 Ethical Challenges for Educators

Teaching ethics is not limited to the transmission of academic content; it also involves a moral responsibility in the comprehensive education of students. Inappropriate behaviors—such as using exams merely as control mechanisms, marginalizing struggling students, refusing to review evaluations, assigning tasks with no educational value, allowing incompetent students to graduate, improperly delegating clinical responsibilities, engaging in harassment or abuse of power, and suppressing critical thinking—are all practices contrary to academic integrity and ethical teaching.

Educators, regardless of the subject they teach, are obliged to respect the dignity of their students, maintain confidentiality, and act with honesty and ethical integrity. This is further reinforced by institutional, professional, and social responsibilities, which demand moving beyond outdated notions—such as the idea that teaching is merely the unidirectional transmission of content or the imposition of personal values.

The success of the educational act is not measured by the brilliance of the teacher, but by the meaningful learning achieved by the student. Reaching this goal in the field of bioethics undoubtedly represents a greater challenge than in other, less complex areas.

14.8 How to Teach Bioethics in a Comprehensive Way

There are two key questions that must be asked—and have been asked—regarding ethical education across different fields: Is it possible to teach ethics? and Is it possible to teach someone to be ethical?

For the first question, and as previously mentioned, it is indeed possible to teach and explain the existing or prevailing ethical norms within a given discipline. However, teaching someone to be ethical tends to be more complex, as this trait is closely linked to the individual's personal development, personality, and various external factors that can affect moral judgment—such as being under pressure. Therefore, it is necessary to design strategies that help students integrate this trait into the reflection of their actions [15].

Because of this, it is important not only to consider the structure of curricula but also to focus on students, teachers, and teaching methodologies throughout the learning process—methods that allow for the development of the skills, attitudes, and knowledge required to understand ethics and to become an ethical person.

While this approach may not significantly impact the acquisition of new concepts—such as current laws or ethical theories—it can help modify the cognitive biases that individuals develop throughout their lives regarding ethical behavior. Herreros [16] points out that observing and sharing specific situations are one of the most effective ways to foster ethical conduct, with leading by example being among the best means to convey values—not through imposition, but through the appreciation of others' attitudes and, when necessary, through dialogue or debate about the behaviors being modeled. In other words, it is not only about teaching what ethics is but also how to live it.

At this point, it becomes clear why the traditional teaching model may not be sufficient in bioethics, which is why the use of different teaching models is proposed to strengthen the necessary skills and attitudes for specific analyses, such as case studies, or more general ones, such as daily activities.

Although there is no definitive model for teaching bioethics, competency-based education and problem-based learning present viable options. In both approaches, the need for students to practice applying theoretical knowledge under the supervision of facilitators, tutors, or mentors is considered. Within this framework, attitudes and skills used and required can be discussed and, finally, assessed to determine if corrections, modifications, or new learning is necessary to face these or new situations and, over time, enable analysis to be performed organically in everyday life [17].

For this reason, and recognizing that learning is a continuous process—where even concepts already learned are fluctuating—and that reflection and abstraction from specific experiences allow the internalization of concepts and theories, which are later evaluated for confirmation or modification, some areas are proposed that should be considered in the ethical training of healthcare professionals in order for it to be regarded as comprehensive:

1. Bioethics and research ethics should be present not only as isolated content but also as a transversal concept throughout undergraduate training. This implies that their guiding principles should be integrated into clinical courses, research activities, community practices, and hospital experiences. Longitudinal inclusion ensures that students do not perceive them as strictly theoretical subjects disconnected from reality, but rather as applicable tools in all phases of their education and professional practice.
2. Informal learning that occurs in the clinical environment, especially through observing the behaviors of professors and professionals, constitutes a fundamental component of ethical training. Often called the hidden curriculum, which can sometimes contradict the formal discourse, it must be recognized and addressed. Comprehensive training in bioethics and research ethics should be accompanied by spaces for critical analysis of observed practices and can help align explicit and implicit messages in the educational process. Incoherence between ethical discourse and the actual practices of educators can undermine the credibility of formal teaching [18].
3. Ethical deliberation requires time and specific conditions. Therefore, didactic spaces should be created for case analysis, handling real or simulated dilemmas, group work supported by faculty, and the development of arguments that connect principles, consequences, and contexts. These spaces should not be reserved solely for bioethics courses but integrated into clinical practices, team meetings, research activities, and interdisciplinary academic sessions.
4. Ethical training involves strengthening critical moral conscience. Encouraging self-criticism allows students to examine their own actions, decisions, and attitudes in relation to the ethical principles learned. This practice fosters the development of a responsible and committed professional identity. Ethical teaching does not impose values but creates conditions for students to discover, question, and authentically assume them.
5. Assessment of ethical competence cannot be limited to written tests or exams that appeal to memory. It is necessary to adopt authentic assessment strategies

that evaluate the ability to deliberate, argue, act with moral sensitivity, and apply ethical principles in real situations. Evaluating bioethics requires methods beyond theoretical exams.

14.9 Integrating Bioethics Education in the Digital Era

Bioethics education must extend beyond formal higher education systems, as healthcare professionals, researchers, and other interested individuals may not always have free access to such training. In this regard, technological advances have enabled the creation of various platforms for self-directed and, in some cases, asynchronous learning, significantly increasing the reach of continuing education efforts by eliminating temporal and geographic limitations [19].

The use of these platforms also provides spaces for discussion among participants, and since there are no geographic barriers, it allows for the comparison of similarities and differences across diverse cultural contexts. This is particularly valuable in regions like Latin America, where shared contexts can broaden perspectives and reduce related biases [19].

Moreover, the benefits are not limited to time and space. These platforms also allow for the development of repositories containing a wide range of sources and perspectives on both common and uncommon bioethical dilemmas, accessible to interested individuals at low or even no cost.

Additionally, other technologies can be explored to create even more immersive learning experiences. Although these may involve higher costs, they enhance interactive learning and support the modeling of consequences in ethical decision-making [20].

14.10 Conclusion

The teaching of bioethics and research ethics in the training of future healthcare professionals must go beyond reductionist approaches that regard it as an optional or secondary subject. Its relevance is comparable to that of clinical competencies, as medical decisions always involve moral judgment. Achieving true ethical performance requires not only theoretical knowledge but also the development of deliberative capacities, reflective attitudes, and a genuine commitment to the well-being of others.

Pedagogical efforts in bioethics must extend throughout the entire curriculum, draw from the hidden curriculum, create systematic spaces for reflection, encourage ethical role-modeling by educators, promote self-criticism, and design evaluations aligned with its educational objectives—while leveraging the innovative technological resources available today, which have expanded greatly since the recent COVID-19 pandemic. Only in this way will it be possible to train healthcare professionals who, in addition to being competent, are also just, compassionate, and responsible in facing the ethical challenges of the twenty-first century, both individually and collectively.

References

1. Pellegrino ED, Thomasma DC. The virtues in medical practice. Oxford University Press; 1993.
2. Pessini L. At the origins of bioethics: from Potter's bioethical creed to Fritz Jahr's bioethical imperative. Revista Bioética. 2013;21(1):9–18.
3. Arellano J, Farías E. Bioética sin biomoral: elementos críticos para comprender la bioética. Secretaría de Salud, Comisión Nacional de Bioética & Universidad Autónoma de Querétaro; 2023.
4. Beauchamp TL, Childress JF. Principles of biomedical ethics. 8th ed. New York: Oxford University Press; 2019. https://doi.org/10.1097/ACM.0000000000004875.
5. Pellegrino ED, Thomasma DC. For the patient's good: the restoration of beneficence in health care. Oxford: Oxford University Press; 2022.
6. Jonsen AR, Siegler M, Winslade WJ. Clinical ethics: a practical approach to ethical decisions in clinical medicine. 9th ed. New York: McGraw-Hill Education; 2022.
7. Macer D. Moral games for teaching bioethics. UNESCO Chair in Bioethics; 2008.
8. Striedinger MP. ¿Qué es formar en bioética? Revista Interamericana de Investigación, Educación y Pedagogía. 2015;8(2):391–411.
9. Outomuro D. Fundamentación de la enseñanza de la bioética en medicina. Acta bioethica. 2008;14(1):19–20. https://doi.org/10.4067/S1726-569X2008000100003.
10. Syrotina O. Bioethical competence as an integral component of future biotechnologists' professional training. Euromentor J. 2023;14(3):88–99. https://doi.org/10.31548/hspedagog14(3).2023.92-98.
11. Morin E. Los siete saberes necesarios para la educación del futuro. Organización de las Naciones Unidas para la Educación, la Ciencia y la Cultura; 1999.
12. Sánchez M. La bioética que vamos a enseñar y el peligro de los reduccionismos éticos. In: Herreros B, Bandrés F, editors. Educar en bioética al profesional de ciencias de la salud. Fundación Tejerina; 2010.
13. Osbeck C, Franck O, Lilja A, Sporre K. Possible competences to be aimed at in ethics education – ethical competences highlighted in educational research journals. J Beliefs Values. 2008;39(2):195–208. https://doi.org/10.1080/13617672.2018.1450807.
14. Vázquez R. Conferencia Magistral en Bioética Juliana González, "Educar en la era digital" [video en Internet]. YouTube; 21 abr 2025. Disponible en: https://www.youtube.com/live/2yULpvPKENs
15. Bazerman M, Tenbrusel A. Blind spots: why we fail to do what's right and what to do about it. Princeton: Princeton University Press; 2011.
16. Herreros B. La educación de los valores profesionales en la medicina. In: Herreros B, Bandrés F, editors. Educar en bioética al profesional de ciencias de la salud. Fundación Tejerina; 2010.
17. Gavriel J. The self-directed learner in medical education: the three pillar model for developing self-directedness. CRC Press; 2017.
18. Hafferty FW, Castellani B. The hidden curriculum: a theory of medical education. In: Wear L, Aultman J, editors. Professionalism in medicine. Cham: Springer; 2020. p. 17–35.
19. Arrieta M, Vidal S, Educación en Bioética a distancia. Una experiencia regional. In: Vidal S, editor. La Educación en Bioética en América Latina y el Caribe: experiencias realizadas y desafíos futuros. UNESCO; 2012.
20. Harfouche A, Nakhle F. Creating bioethics distance learning through virtual reality. Cell Press Rev. 2020;38(11):1187–92.

Chapter 15
Bioethical Education in Ibero-America: Main Characteristics and Current Challenges

Alejandro Sánchez-Guerrero, María Victoria Fernández Molina, and Daniela Casillas-Gómez

15.1 Introduction

Warren Reich defined bioethics as "the systematic study of human conduct in the life sciences and health, as this conduct is examined in light of moral values and principles" [1]. This definition highlights the importance of bioethics as a field that connects life sciences and health with moral principles, aiming to address complex issues [1]. André Hellegers introduced bioethics into academic and biomedical domains [1], recognizing the need for a discipline to address ethical dilemmas in these fields. This development led to the establishment of the Kennedy Institute of Bioethics at Georgetown University in 1971, marking the start of bioethics as an academic discipline [2]. Under the leadership of Edmund D. Pellegrino [2], this institute excelled in exploring ethical dilemmas in medicine, biology, and public health.

The Kennedy Institute distinguished itself through its commitment to bioethical education, offering programs for students and healthcare professionals that addressed topics such as informed consent, end-of-life decisions, resource allocation in healthcare, and ethics in human research. Since its founding, bioethics has expanded its scope globally and across various areas of specialization. In Venezuela, for example, bioethics was introduced in the 1980s, following Hellegers' approach, with a particular emphasis on health sciences [1]. Subsequently, it consolidated in

A. Sánchez-Guerrero (✉) · D. Casillas-Gómez
Centro Anáhuac de Desarrollo Estratégico en Bioética (CADEBI), Universidad Anáhuac México, Mexico City, Mexico
e-mail: alejandro.sanchezg@anahuac.mx; daniela.casillasgo@anahuac.mx

M. V. F. Molina
Inter-American Academy of Human Rights (AIDH), Universidad Autónoma de Coahuila (UAdeC), Saltillo, Mexico
e-mail: victoria.fernandez@uadec.edu.mx

D. Cerdio-Domínguez, M. E. de los Ríos-Uriarte (eds.), *Current Bioethics*, New Paradigms in Healthcare, https://doi.org/10.1007/978-3-032-25294-4_15

other sciences and fields of knowledge, seeking reconciliation of human knowledge so that science and the humanities "walk in the same direction and sense" [1]. Bioethics encompasses more than just interpersonal ethics; it adopts a broader and more global perspective, emphasizing the belief that human life is deeply interconnected with that of other living beings [1].

Since the establishment of the Kennedy Institute, bioethical education has experienced global expansion and development, driven by the discipline's relevance in protecting human dignity within biomedicine. Among the most notable areas requiring specialized professionals are: first, academic research on ethics in medicine, biotechnology, and biomedical research; second, teaching and training in educational institutions such as universities, medical schools, and postgraduate health programs, covering various topics related to bioethics; third, ethical consulting, which involves providing advice to health institutions, ethics committees, researchers, and clinical professionals; fourth, clinical ethics, which evolves in collaboration with healthcare teams to address ethical issues in patient care; fifth, the development of public policies related to health and biotechnology; and finally, participation in research ethics committees, which are responsible for reviewing protocols to ensure adherence to ethical principles and the protection of participants' rights and well-being in studies [3].

The United Nations Educational, Scientific, and Cultural Organization (UNESCO) advocates that humanistic professional training in health should be a priority in university education for the twenty-first century [4]. In this context, ethics and bioethics disciplines in medical curricula are essential for preparing graduates to attend to patients in their biopsychosocial and spiritual dimensions [4]. Therefore, incorporating these subjects at the undergraduate level is crucial for future professionals to develop skills that allow them to establish a human relationship of mutual respect with patients [4]. Contemporary society no longer accepts the asymmetrical relationship between healthcare professionals and patients, known as medical paternalism, and individuals demand the ability to express their autonomous will [4].

The integration of bioethics into health curricula is essential. Some proposals include civil ethics and a clinical ethics approach across various curricular components, such as communication with patients and their families [5]. Specific bioethics courses and transversal approaches in medical programs have also been developed [5]. There is a growing consensus on the vertical and horizontal integration of ethical training throughout the medical educational process, which encompasses both preclinical and clinical periods [5].

However, the teaching of bioethics faces various challenges and limitations. For example, a study on the teaching of ethics and bioethics (EEB) in the Multiprofessional Health Residency Programs (PRMS) at the Universidad Federal de Rio Grande do Norte (UFRN) in Brazil revealed that the pedagogical projects offered fewer hours and thematic modules than those recommended by UNESCO in its "Core Curriculum" [6]. Additionally, it was found that the preceptors were unaware of EEB's inclusion in the PRMS curriculum. This study also highlighted communication and epistemological failures that limit the approach to bioethics in both training and health practice [6]. The time explicitly dedicated to ethics and bioethics in some residency programs is minimal, representing only a small percentage of the total teaching hours [6].

15.1.1 Bioethics Education in Latin America

Bioethics was introduced in Latin America and Europe through the globalization of ethical debates from the United States. Despite its American origins, bioethics in Latin America developed a distinctive character, adapting to the local realities of the region, particularly in relation to human dignity and social justice. In Argentina, José Alberto Mainetti championed a shift toward more humanistic medicine. In the 1980s, he founded the Escuela Latinoamericana de Bioética (ELABE), which has played a crucial role in bioethics education in the region. Similarly, in 1985, Fernando Sánchez Torres established the Instituto Colombiano de Estudios Bioéticos (ICEB) in Colombia, an institution that has maintained a continuous seminar on bioethics [7].

In a study on bioethical issues and the evolution of moral development in medical students, Diana C. Sierra Trujillo [2] identified that various Colombian universities have incorporated bioethics courses with diverse approaches, reflecting the array of ethical perspectives and concerns in the field of medicine. Among the universities mentioned, the Universidad de los Andes has implemented courses focused on medical ethics and professionalism. Simultaneously, the Pontificia Universidad Javeriana emphasizes the doctor–patient relationship and the medical act. Concurrently, the Universidad El Bosque offers specialized seminars in bioethics, while the Universidad del Rosario promotes training in ethics and values, along with ethics and medical law. This variety of approaches underscores the commitment of Colombian academic institutions to integrate bioethics into the training of future health professionals, acknowledging the importance of addressing not only the technical aspects of medicine but also the ethical challenges that arise in professional practice.

According to Cardona Ramírez [7], the adaptation of bioethics in Latin America responds to the need to contextualize it within the region's social, cultural, and political realities. In contrast to the individualistic focus of bioethics in the United States and Europe, Latin American bioethics has adopted a perspective oriented toward social justice and equity, addressing issues such as the right to health, poverty reduction, and respect for cultural diversity. This holistic approach extends beyond the doctor–patient relationship to encompass social and economic rights, ensuring a dignified and equitable life for the entire population.

Bioethics in Ibero-America began to develop more robustly in the 1990s, influenced by advances in medicine and biotechnology, along with the necessity to address ethical issues within these fields. It is noteworthy that educational institutions and healthcare professionals have collaborated to establish programs aimed at tackling these challenges in research and clinical practice. Among the most relevant programs from various countries in the region are the following:

In Spain, which boasts a diverse range of academic training in bioethics, there is an online doctorate in Bioethics and Law at the Universidad de Barcelona. Its primary goal is to provide advanced training to researchers in various fields of bioethics, in relation to Human Rights, from an interdisciplinary perspective. Similarly, the doctoral program in Health, Well-being, and Bioethics at the Institut Borja de Bioética, Universitat Ramon Llull, focuses on equipping students with the knowledge and skills necessary to develop innovative and creative research models that address the social and healthcare demands of the twenty-first century in an interconnected manner.

The increasing significance of bioethical reflection has resulted in a greater availability of specialized educational programs in this field. Traditionally, these programs have concentrated on theoretical teaching models, descriptive analyses, and the examination of practical cases, reflecting the dominant perspectives and prevailing argumentative models in bioethics [8]. In this context, many students who choose to pursue studies in bioethics aim to gain the necessary training to integrate into hospital ethics consultation committees or research ethics committees, which are becoming increasingly common in healthcare and research institutions in Spain [8].

However, Spain is experiencing a significant debate about the necessity of addressing the limitations inherent in purely theoretical or casuistic teaching models. In this regard, there are calls for implementing comprehensive training programs in bioethics that include case studies and ethical theories, while integrating an interdisciplinary approach. Such training should combine a solid foundation in anthropology and ethics, offering a broad perspective on life and human dignity [8].

In response to this need, bioethics education in Spain has evolved, primarily focusing on postgraduate levels. This strategy has shown a trend toward more comprehensive education that merges theory with practice and is rooted in disciplines such as anthropology and ethics. The goal is to prepare competent professionals who can address the complex bioethical dilemmas that arise in various fields, in both healthcare and research contexts, and who deeply understand the ethical, social, and cultural dimensions underlying these dilemmas.

In Mexico, the development of bioethics has been remarkable, especially in educational offerings. Two institutions stand out in teaching this discipline: first, the Universidad Nacional Autónoma de México (UNAM), which offers master's and doctorate programs in bioethics that provide advanced training focusing on clinical ethics, research ethics, and public health policies. Second, Universidad Anáhuac, through its Faculty of Bioethics, offers a doctorate and a master's in bioethics that address ethical issues in medical practice and biomedical research, emphasizing a human dignity–centered perspective.

In Mexico, educational offerings in bioethics have experienced significant development, driven by the need to train human resources that address social demands and contribute to collective well-being [9]. Various higher education institutions, such as the Universidad Nacional Autónoma de México (UNAM) and Universidad Anáhuac, excel in this field [10]. UNAM offers master's and doctorate programs in bioethics with an advanced focus that encompasses clinical ethics, research ethics, and public health policies. Meanwhile, Universidad Anáhuac, through its Faculty of Bioethics, also delivers comprehensive training in master's and doctorate programs, focusing on ethical issues related to medical practice and biomedical research, always from a perspective centered on human dignity. Additionally, several other institutions, such as the Bioethics Unit at the Universidad Autónoma de Querétaro and the Centro de Investigación Social Avanzada, enhance this educational offering, which includes programs at various levels, from specializations to doctorates, with a strong emphasis on ethical approaches committed to human rights and social issues. Bioethics in Latin America, and by extension in Mexico, has gone through stages of reception, assimilation, and recreation, striving for its own identity [11].

An enhanced educational context has also accompanied the development of bioethics in Mexico, supported by the use of digital resources and innovative methods such as case discussions, backed by Redbioética UNESCO [12]. However, structural challenges persist despite these advances, leading to educational modalities that cater to specific interests. Mexico has actively participated in initiatives and collaborations at regional and international levels, promoting academic exchange and ethical reflection.

In Argentina, there has been significant progress in bioethics training. Notable in this area are the master's in bioethics from the University of Buenos Aires (UBA), an interdisciplinary program that focuses on medical ethics, research ethics, and applied bioethics, and the master's program in bioethics at the Facultad Latinoamericana de Ciencias Sociales (FLACSO Argentina), which addresses ethical issues in biomedicine and public health.

Brazil has consolidated bioethics from a perspective focused on social justice and human rights, while also maintaining influences from American bioethics in the medical field. In this context, the doctorate program in Bioethics, Applied Ethics, and Collective Health at the Universidade Federal do Rio de Janeiro stands out, with the primary objective of advancing academic development in bioethics and research ethics. Additionally, the Universidad de Brasilia (UnB) offers a master's and a doctorate in bioethics with an emphasis on social justice and human rights, aimed at exploring and expanding knowledge of bioethics concerning the prevalent issues in Brazil and Latin America, while also providing multidisciplinary training for professionals in various related fields.

The establishment of the Sociedade Brasileira de Bioética in the 1990s marked a significant milestone in the development of bioethics in Latin America, consolidating itself as a turning point in the region [6]. This event highlighted the growing importance of bioethics not only as an academic discipline but also as a crucial tool for addressing ethical challenges arising from social and economic inequality. In this framework, the so-called "intervention bioethics" has emerged, an approach that seeks to overcome social disparities and assume collective responsibility in the face of bioethical issues, especially in contexts of vulnerability. However, despite advancements in establishing this discipline, the teaching of bioethics in Brazil remains deeply influenced by the principles of the principlism model, which is prevalent in Anglo-American traditions, and continues to center on the biomedical field, limiting the integration of broader and more social perspectives [6]. In South America, bioethical issues are more complex and wider than in high-income countries of the northern hemisphere. These problems encompass not only biomedical and clinical aspects but also social, economic, environmental, and public health issues, reflecting the plurality of lived realities among the region's populations [8]. This diversity of context underscores the need for a bioethical approach that is not limited to medicine or biotechnology but also addresses the challenges arising from structural inequalities and public health policies.

In this sense, it is essential to recognize the different perspectives for analyzing bioethics developed in various cultures. While the Anglo-American tradition primarily relies on the principlist approach, which focuses on the principles of autonomy, beneficence, non-maleficence, and justice, European currents tend to emphasize the social dimension of bioethics, highlighting the relevance of the principle of justice and its application in the equitable distribution of resources and healthcare [8]. This contrast underscores the need for critical reflection on how various cultural and philosophical traditions influence bioethical policies, suggesting that truly global bioethics must incorporate a plurality of perspectives to address ethical health issues in a more inclusive and contextualized manner.

Chile has also distinguished itself in developing bioethics, showcasing a remarkable educational offering. Among its institutions, the master's in social bioethics and public health at Universidad Central de Chile stands out, focusing on preparing health professionals to identify ethical criteria in clinical decisions, recognize the human aspects of care, and provide compassionate treatment to users of the health system. Another notable program is the master's in bioethics and its current challenges, offered by Pontificia Universidad Católica de Chile, which addresses ethical aspects of related areas such as research ethics, social health, technology, and the environment.

Several Chilean universities, including the Universidad Regional de O'Higgins (UOH) and the Universidad de Talca, have undertaken substantial curricular transformation processes aimed at incorporating ethics and bioethics into their students' professional training. This trend reflects an increasing awareness of the necessity to address ethical aspects in the health sciences field, fostering a more holistic education that considers the moral implications of professional practices. The UOH experience, developed from 2017 to 2022, serves as a representative example of these efforts, characterized by the design and implementation of ethics and bioethics courses following a competency-based approach, particularly focused on health science programs [13].

These courses, designed from a pedagogical perspective that prioritizes competency development, aim not only to provide a theoretical framework but also to cultivate in students a profound understanding of the value factors influencing their professional practice. Furthermore, these programs focus on examining the impact of scientific and technical advancements on moral and ethical relationships, emphasizing the importance of reflecting on how progress in medicine and biotechnology can lead to complex ethical dilemmas. Additionally, they seek to equip students to effectively identify and tackle the frequent ethical dilemmas encountered in healthcare professional practice [13]. This initiative underscores the need to train health professionals who are not only technically knowledgeable but also possess a solid ethical foundation, enabling them to face the challenges posed by the contemporary biomedical landscape.

Similar to other regions, bioethics education in Ibero-America is characterized by its interdisciplinary approach, integrating perspectives from medicine, philosophy, law, and social sciences. The applied ethics methodologies in clinical and research contexts, the casuistry tailored to the cultural and legal frameworks of each country, and international collaborations to enhance programs and promote knowledge exchange serve as fundamental pillars in developing bioethics in the region.

Various methodologies for teaching bioethics have been researched, highlighting those that promote an active approach and foster students' critical development [14]. Among the most relevant teaching techniques is the case study method, which is widely used in the region as an effective pedagogical tool [15]. Additionally, narrative ethics emerges as a valuable methodology for bioethical deliberation, employing literary works and their staged representations to analyze ethical conflicts, encouraging deep discussion and reflection on these dilemmas [16].

In general terms, academic training in bioethics in Ibero-America, particularly in Latin American countries, is undergoing a process of development and consolidation. This process involves integrating bioethics into health curricula, exploring effective teaching methodologies, and addressing the ethical and moral dimensions inherent in professional practice. Despite these advancements, various challenges have emerged, particularly the need for greater practical engagement of students with real bioethical dilemmas and the ongoing training of educators in this discipline. These challenges emphasize the importance of enhancing bioethics education in theoretical terms and its practical application to ensure that future health professionals have the tools to tackle the complex ethical problems that arise in their professional practice.

UNESCO has been crucial in promoting and educating bioethics worldwide, especially in Latin America and the Caribbean. Its support has established a strong foundation of knowledge, values, and specific approaches considering the region's cultural, historical, and social contexts. UNESCO has encouraged the development of ethical reflection that responds not only to scientific and technological advancements but also to the unique circumstances of the people in Latin America and the Caribbean.

One of UNESCO's main contributions in this field, alongside the Universidad Autónoma de México, has been the publication of the manuals "Volume I: Bioethics Education Manual—The curricular agenda in bioethics: opening horizons" and "Volume II: Bioethics Education Manual—Grounding bioethics: Knowledge, values, and visions from Latin America and the Caribbean." These manuals provide a comprehensive perspective on the most relevant bioethical issues for the region, promoting critical reflection on the relationship between scientific advancements, medical practices, human rights, and social issues. The manuals also include approaches that link bioethics with topics such as health equity, social justice, cultural diversity, and sustainability.

Volume I explores key issues affecting bioethics in Latin America and the Caribbean, including ethical dilemmas in biomedical research, healthcare, genetics, and the protection of patient rights, all framed by the sociopolitical realities of the region. In this way, not only are universal concepts of bioethics examined, but they are also contextualized within local particularities, such as ethnic and cultural diversity, poverty, disparities in access to health services, and the impacts of economic development policies [11].

Conversely, Volume II centers on education in bioethics, a vital aspect of consolidating the discipline in the region. This manual is dedicated to training health professionals, educators, students, and citizens, offering pedagogical tools and methodological approaches aimed at integrating bioethics into educational programs and public policies. Through its content, it seeks to strengthen the knowledge, values, and visions that will guide future generations in making responsible ethical decisions in both professional arenas and the social and political context [12].

UNESCO, with its ongoing commitment to bioethics education, aims not only to enhance the quality of debate on these issues in Latin America and the Caribbean but also to help foster a culture of respect, human dignity, and social justice. The manuals published by UNESCO have been essential for health professionals, policymakers, and society in addressing bioethical challenges reflectively, inclusively, and sustainably, thereby contributing to more equitable human development in the region.

Currently, bioethics in Ibero-America has established itself as a fundamental discipline oriented toward social justice and human rights, addressing the ethical dilemmas arising from biomedical development. This approach has led to the creation of academic programs and institutions that train professionals to apply bioethical principles in various contexts, underscoring the increasing significance of this discipline in the education of future doctors and other health professionals. In this sense, bioethics has become a crucial tool for tackling the ethical challenges in medical practice, promoting a more humane and responsible approach toward patients and society. However, a relevant finding in the work of Mayara Valeska Torres Mendoza [17] reveals that, in Latin America and the Caribbean, only 53% of medical schools (276 out of 517) include the subject of bioethics in their academic curriculum. Brazil stands out as the country that has achieved the greatest institutionalization of this subject and its evaluation, distinguishing itself for its efforts to integrate and measure the outcomes of bioethics in medical training. This highlights a significant challenge for the region: fully institutionalizing bioethics within medical education. Furthermore, another notable challenge arises in evaluating behavioral changes in future doctors who receive training in bioethics through various modalities. The lack of systematic research in this field creates a critical gap in understanding the real effects of bioethics education on the professional and ethical development of medical students, underscoring the urgent need to advance in this area of research [17].

15.1.2 Bioethical Training in Response to the Needs of Hospital and Research Committees

Bioethics in Ibero-America has emerged as a key discipline for reflecting on the ethical dilemmas arising from medicine, biomedical research, and the physician–patient relationship, with a strong emphasis on human dignity and social justice. In this context, hospital and research ethics committees play a fundamental role in advising healthcare professionals on complex situations, such as end-of-life decision-making, pain management, and the protection of patients' rights. These committees have promoted comprehensive ethical training that adapts to the cultural, religious, and socioeconomic contexts of each country, ensuring equity in medical care and respect for patients' autonomy [10].

Bioethics committees in Latin America have been crucial for institutionalizing bioethics within hospital settings, with countries like Mexico, Brazil, and Argentina leading the way. In Mexico, the establishment of the Comisión Nacional de Bioética (CONBIOÉTICA) and the enactment of the Ley General de Salud (General Health Law) facilitated the creation of bioethics committees in hospitals, fostering bioethical training for healthcare professionals. In Brazil and Argentina, hospital bioethics committees emerged in the 1990s and 1980s, respectively, to address the ethical dilemmas arising from technological advancements in medicine, promoting interdisciplinary deliberation and the dignity of patients in contexts of inequality [18, 19].

In the region, research ethics committees have developed to oversee biomedical research and protect participants, especially those in vulnerable situations. Since the 1990s, the increasing participation of Latin America in international clinical trials has driven the creation of these committees, which review research protocols and ensure compliance with fundamental bioethical principles, such as autonomy and justice. Mexico, Brazil, and Argentina have taken the lead in establishing these committees, focusing particularly on protecting vulnerable populations and promoting research ethics that prevent exploitation [18].

Despite advances, the consolidation of bioethics in Latin America faces challenges, particularly regarding academic training. Strengthening bioethics education at both undergraduate and graduate levels has become a strategic focus to ensure the effectiveness and sustainability of bioethics committees. In countries such as Mexico, Colombia, and Brazil, initiatives have been launched to connect bioethics committees with universities and research centers, promoting diploma programs, master's degrees, and specialized courses. These initiatives aim to train future committee members, equipping them to tackle the complex ethical dilemmas that arise in the context of medical care and research [10]. Academic training in bioethics serves as a transformative tool that encourages more inclusive and equitable ethical practices. It enables bioethics and research ethics committees to establish themselves as benchmarks for advancing social justice and human dignity in the region. Institutions like UNESCO have also significantly contributed to training professionals through workshops and collaborative networks that reinforce bioethics in vulnerable contexts. These initiatives not only address current ethical needs but also help build regional ethical leadership capable of proposing innovative and culturally sensitive solutions in the medical and scientific fields [18].

15.2 Methodology

The purpose of this work is to present the results of a systematic exploration of academic offerings in bioethics across Ibero-American countries. This information is part of the *Atlas Interactivo Iberoamericano de Bioética* [20], developed by the Centro Anáhuac de Desarrollo Estratégico en Bioética (CADEBI) at Universidad Anáhuac Mexico.

15.2.1 Data Collection Strategy

The identification of the academic offerings was conducted through a 16-month search strategy (from July 2023 to November 2024), utilizing various online information sources across Ibero-American countries. The employed strategy included:

- Search Engines: Continuous monitoring was performed using search engines such as Google, Bing, and Yahoo.
- Academic Databases: Articles published between 2019 and 2023 were reviewed on platforms like Google Scholar, SciELO, Redalyc, PubMed, Scopus, and Web of Science to identify the institutions linked to authors in the field of bioethics.
- Specialized Institutions and Organizations: Institutions, centers, associations, and organizations, both national and international, listed in the *Atlas Interactivo Iberoamericano de Bioética* were consulted to determine their educational offerings in bioethics (courses, diplomas, specialties, master's degrees, doctorates, and postdoctoral programs).
- Other Sources: Specialized bioethics journals, academic social networks like LinkedIn and Academia.edu, and social media platforms such as Facebook, Instagram, YouTube, X (Twitter), Pinterest, and Snapchat were also considered.

15.2.2 Inclusion Criteria and Evaluation

In cases where information sources were limited or scarce, the relevance of the institution and the historical significance of its academic offerings in the field of bioethics, both nationally and internationally, were evaluated. All identified academic offerings were considered an inclusion criterion, regardless of their status at the time of the research.

15.2.3 Coding and Analyzed Variables

From the collected information, several variables were established to classify the academic offerings, and coding was executed in the database. The variables included:

- Educational Level: Doctorates, master's degrees, specialties, diplomas, seminars, and courses.
- Subsequently, a recording of the educational level options was conducted; a new variable called Training Trends was established. Records of doctorates and master's degrees were classified under "academic trend," while records of specialties, diplomas, and seminars or courses were classified under "professional trend."
- Country of Offering: The country where the program is offered.
- Modality: In-person, online, or hybrid.
- Field of Knowledge: Clinical (clinical bioethics, dilemmas in medicine, scientific research), socio-legal (human rights, law, health justice), and environmental (environmental ethics, environmental justice, global bioethics).
- Institutional Linkage: Collaboration among institutions in educational offerings, whether present or absent.
- Language of Instruction.
- Type of Institution: Universities, associations, institutes, organizations, and national or state commissions.
- Graduate Profile: Research, professionalization, and support for hospital bioethics and research ethics committees.
- Status of Offering: Current (last year), partially current (the previous two years), and not current (more than 2 years ago).

15.2.4 Data Analysis

The resulting database was analyzed using SPSS (Statistical Package for the Social Sciences) software. The following statistical tests were conducted:

- Contingency Tables: To examine the frequency distribution of categorical variables.
- Chi-Square Test: To assess the statistical significance of the relationship between two or more categorical variables.
- Pearson's R Coefficient: To evaluate the likelihood of the variables being related and dependent on one another.

15.3 Results

The systematic exploration of academic offerings in bioethics throughout Ibero-America, conducted through an extensive search and data collection strategy, has allowed for the mapping and characterizing of training options available in the region. The main results for each analyzed variable are presented below:

15.3.1 *Educational Offerings in Bioethics in Ibero-America*

The academic offerings in bioethics in Ibero-America show a significant concentration, with 83 courses and 76 master's programs accounting for 31.8% and 29.1% of the total offerings, respectively. A total of 65 diplomas or seminars were identified, making up 24.9%, while 22 specializations and 15 doctoral programs are less prevalent, representing only 8.4% and 5.7%. This distribution, clearly illustrated in Fig. 15.1, indicates a preference for short- or intermediate-duration programs that are accessible to a broad audience seeking training in bioethics, although opportunities for deeper levels of specialization in the region are limited.

15.3.2 *Educational Modalities*

Regarding the training modalities, doctoral and master's programs are primarily in-person, with proportions of 60% and 52%, respectively. In contrast, specializations show a marked trend toward the online modality, with 61.9% of programs offered in this format. Diplomas and courses also favor online modality, with 68.5% and

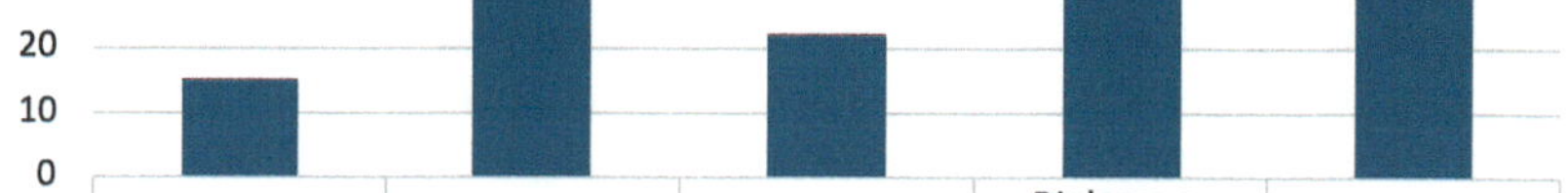

	Doctorate	Master's Degree	Specialization	Diploma or Seminar	Course
Percentage	5.70%	29.10%	8.40%	24.90%	31.80%
Number	15	76	22	65	83

Fig. 15.1 Bioethics educational offerings in Ibero-America by educational level

55.2%, respectively. This trend indicates that the online modality is preferred for short-duration training programs due to its accessibility and lower demand for physical resources.

The Chi-square test reveals a statistically significant relationship between program modality (in-person or online) and educational level ($p < .001$). This suggests that the distribution of modalities is not random concerning the type of academic program; instead, there is a tendency that connects them. In other words, certain types of programs are more likely to be delivered in specific modalities, as demonstrated by the prevalence of the online modality in short-duration programs such as diplomas and courses.

On the other hand, the Pearson correlation coefficient ($R = .143$) reveals a positive correlation between the variables, although this correlation is weak. A positive correlation indicates that as the educational level of the programs increases, there is a slight tendency for the modality to shift. However, since the correlation is weak, the relationship is neither strict nor linear; it merely suggests that there is a slight association, without being deterministic, between the program's level and its modality.

15.3.3 Program Status

When assessing the status of these programs, detailed in Fig. 15.2, it is clear that the online modality has a higher proportion of active programs, particularly at the diploma and course levels, representing 53% of the total academic offerings. This suggests that online programs remain active more frequently than in-person and hybrid programs, likely due to their adaptability and ease of updating. In contrast, in-person programs show a lower proportion of active status, emphasizing higher

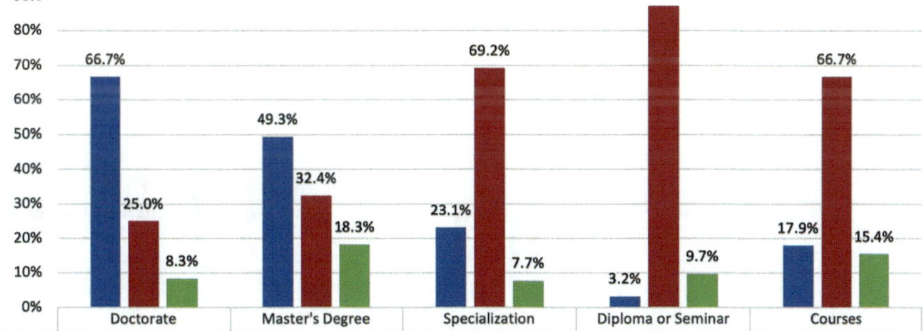

	Doctorate	Master's Degree	Specialization	Diploma or Seminar	Courses
In-person	66.7%	49.3%	23.1%	3.2%	17.9%
Online	25.0%	32.4%	69.2%	87.1%	66.7%
Hybrid	8.3%	18.3%	7.7%	9.7%	15.4%

Fig. 15.2 Current educational level and modality of bioethics programs

academic levels such as master's and doctoral programs, where stability and continuity appear more consistent. The hybrid modality, however, exhibits lower frequency and active status, possibly due to the difficulties of coordinating both in-person and online sessions, which affects their long-term sustainability.

The Chi-square test ($\chi^2 = 39.768$, $p < 0.001$) confirms that a statistically significant relationship exists between the modality of the programs (online, in-person, or hybrid), educational level, and the active status of these offerings. This indicates that whether a program is currently operational is closely linked to both its modality and academic level.

In other words, online programs tend to be more appealing, especially at the diploma and course levels, largely due to their flexibility and ease of updating. Conversely, while less frequently available, in-person programs are more common at higher academic levels, where stability is more pronounced. The hybrid modality has even less ongoing status, possibly because of challenges in coordinating in-person and online sessions, which could affect their sustainability.

15.3.4 Countries with the Most Active Educational Programs

Figure 15.3 visually represents the distribution of active educational programs by country, revealing Mexico's broad diversity across all levels, with substantial representation in each category. Spain and Argentina also offer a significant range of levels, particularly in master's programs, specializations, and diplomas. Brazil and Colombia offer a more concentrated selection, mainly at the master's and specialization levels, although they have less diversity in other areas.

The Chi-square result (61.697, df = 28, $p < .001$) indicates a significant association among the country offering the programs, the educational level, and the active status of the academic offerings. This value suggests that some countries favor certain educational levels and that the availability of these programs varies by country.

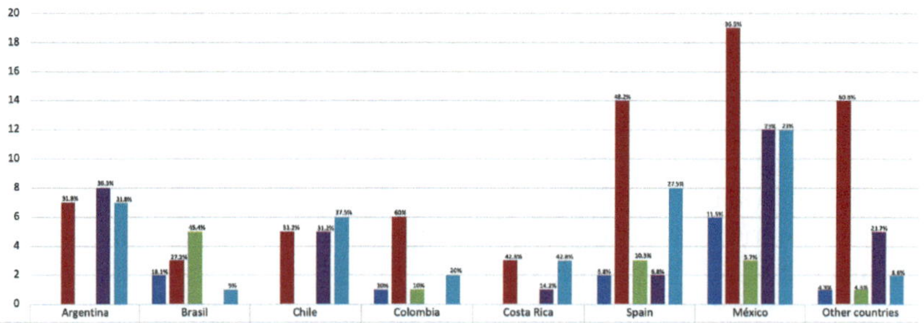

	Argentina	Brasil	Chile	Colombia	Costa Rica	Spain	México	Other countries
Doctorate	0	2	0	1	0	2	6	1
Master's Degree	7	3	5	6	3	14	19	14
Specialization	0	5	0	1	0	3	3	1
Diploma or Seminar	8	0	5	0	1	2	12	5
Course	7	1	6	2	3	8	12	2

Fig. 15.3 Countries with the largest current academic offerings in bioethics

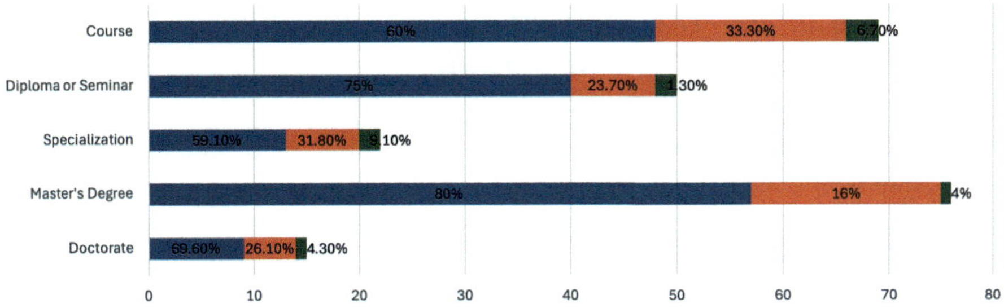

	Doctorate	Master's Degree	Specialization	Diploma or Seminar	Course
■ Clinical	9	57	13	40	48
■ Socio-legal	5	18	7	8	18
■ Environmental	1	1	2	2	3

Fig. 15.4 Distribution of areas of knowledge in bioethics programs

15.3.5 Areas of Knowledge in Bioethics

As depicted in Fig. 15.4, the analysis of knowledge areas in bioethics reveals a clear dominance of clinical topics across all educational levels, particularly in advanced programs such as master's and doctoral degrees, making up between 60% and 75% of the content, which reflects a practical focus in the healthcare sector. In contrast, socio-legal topics have moderate representation, ranging from 23.7% to 33.3%, while environmental issues exhibit low representation overall, with a maximum of 9.1% in specializations and a minimum of 1.3% in master's programs.

The Chi-square test results ($p = 0.523$) and Pearson's R (-0.001, $p = 0.989$) suggest no significant relationship between the area of knowledge and the academic level. Clinical topics prevail across all educational levels, especially advanced ones, while socio-legal and environmental areas have less engagement. This indicates that the selection of areas of knowledge is driven more by personal preferences within the field of study than by the academic level itself.

15.3.6 Graduate Profile

When it comes to graduate profiles, higher academic levels, such as doctorates and master's degrees, generally emphasize research, while specialty levels and shorter programs (like diplomas, seminars, and courses) are typically more practical or supportive. This relationship aligns with the statistical results of the chi-square test ($\chi^2 = 88.511$, $p < .001$), which indicates a statistically significant relationship between the graduate profile and educational level, confirming that these variables are interconnected. Furthermore, the Pearson correlation coefficient ($R = .643$, $p < .001$) suggests a moderate-to-high positive correlation, indicating that as educational levels rise, there is a tendency toward graduate profiles that are more research-oriented, while lower levels lean toward supportive or practical application profiles.

15.3.7 Training Trends in Countries with the Highest Educational Offer in Bioethics

Figure 15.5 provides a detailed overview of the training trends in bioethics across Ibero-American countries, demonstrating a predominant leaning toward a professional orientation (64.9%). This trend is particularly evident in nations such as Argentina, Mexico, and Chile. In contrast, Spain displays a more balanced distribution, with a higher proportion of academic training (35.1%).

The Chi-square analysis indicates a significant relationship between the country and the tendency toward academic or professional training ($p = 0.017$), suggesting that these trends are not uniformly distributed across nations. This implies that the training approach varies according to the national context, potentially reflecting differences in local labor market demands or educational policies.

Meanwhile, the Pearson correlation shows an R value of -0.100 (with $p = 0.106$), indicating a weak and non-significant negative correlation between the country and the type of training trend. This suggests that while differences in training trends by country exist, the relationship does not suggest thet there is a consistent association between the variables.

In relation to the academic trend in bioethics, which comprises 35.1% of the offerings in Ibero-America, some countries stand out by exceeding this average. Spain leads with 57.6% of its programs focused on academic training, followed by Brazil at 47.1%. These countries display a particular interest in a more academic approach to bioethics compared to the regional average, sharply contrasting with other nations like Argentina, Chile, and Mexico, where professional training predominates significantly.

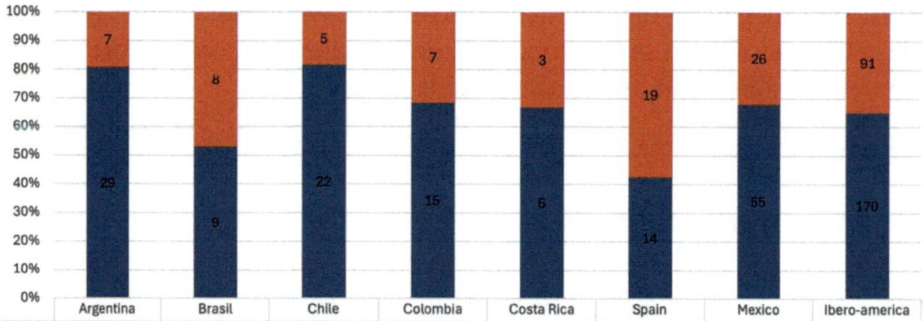

	Argentina	Brasil	Chile	Colombia	Costa Rica	Spain	Mexico	Ibero-america
Academic Trend %	19.4%	47.1%	18.5%	31.8%	33.3%	57.6%	32.1%	35.10%
Professional Trend %	80.6%	52.9%	81.5%	68.2%	66.7%	42.4%	67.9%	64.90%

Fig. 15.5 Training trends in countries with the largest educational offerings in bioethics

15.3.8 *Offer Language*

The language of instruction in the academic offerings in bioethics is predominantly Spanish, especially at the course and diploma levels, where it represents 90.3% and 100%, respectively. Portuguese is also present, though in smaller proportions: 4 doctorates, 5 master's degrees, 7 specializations, and 3 courses are offered in countries such as Portugal and Brazil. In contrast, English has an extremely limited presence, with only two master's programs taught exclusively in English: the Master of Science in Bioethics from St. George's University in Granada and the Master in Global Bioethics, an M.S., at Universidad Anáhuac Mexico. The limited offerings in English could restrict the visibility and collaboration of Ibero-American bioethics programs in global academic networks, diminishing access for non-Spanish-speaking students and the possibility of establishing international connections in bioethics. Results from the Chi-square analysis ($p < .001$) indicate a statistically significant relationship between language and academic level.

The variables Institutional Linkage and Institution in Charge of the Offer were not included in the analysis, as they did not provide relevant information.

15.4 Discussion

Of the five categories of academic offerings (doctorate, master's, specialization, diploma, and course), the most common are courses, followed by diplomas and master's programs. This trend may be a result of the high demand for bioethical knowledge in the region and how this need has been addressed.

With a greater number of short-duration programs available, it is possible to reach more people in a more practical and timely manner. However, mid-to-high-level knowledge, such as master's programs, occupies a privileged position. As the third most common offering, we can infer that there is a thirst for deeper knowledge of bioethics in Ibero-American countries, and an effort is being made to provide the necessary knowledge.

Less significantly, we find specialization in bioethics, which may relate to the fact that these programs are often designed for very specific areas, and not all regions have sufficiently developed labor markets to justify their offering. Additionally, the Ibero-American educational system has largely adopted the Anglo-Saxon model, where specialization is typically achieved through master's programs or technical certifications. On the other hand, access to doctorates is more challenging, resulting in a lower frequency of their delivery.

Furthermore, the modality of offering (in-person, online, or hybrid) plays a fundamental role in understanding the behavior of academic offerings. With over half of the courses and diplomas available online, the convenience and lower economic costs serve as an anchor that allows these offerings to stand out from the "traditional" ones, which are primarily in-person, such as master's or doctorate programs. Conversely, both master's and doctorate programs still largely occur in an in-person format, as advanced learning benefits from in-depth discussions, collaborative problem-solving, and practical activities that are better facilitated through face-to-face interaction. However, despite the predominance of in-person master's and doctorate programs, there is an increasing number of online options, especially in theoretical disciplines, due to technological advancements and the growing acceptance of distance learning. Regarding educational orientation, it is observed that the clinical approach predominates at most educational levels, reflecting the centrality of ethical dilemmas in the health field. Nonetheless, there is underrepresentation in environmental topics, suggesting an opportunity to expand academic offerings in this area, given the increasing interest in environmental ethics and its relevance today. The prioritization of clinical training may be linked to its historical origins, a greater demand in health for this topic, limited specialized resources in environmental ethics, and a public perception that assigns less urgency to environmental dilemmas.

On the other hand, the trend toward professionalization in several countries highlights a practical approach to applying bioethics in clinical and hospital contexts, addressing the immediate need for trained professionals capable of resolving ethical dilemmas in specific work environments. However, some countries, such as Spain and Brazil, exhibit a preference for a more academic training approach, indicating the necessity for a balanced strategy that considers both academic preparation and practical professionalization, promoting comprehensive training that incorporates theoretical knowledge as well as its application in real contexts.

In terms of relevance, online programs show a higher level of continuity, which may be attributed to their ease of updating and reduced dependence on physical resources. However, in-person programs at higher levels of education tend to maintain relevance for a longer period, possibly due to their stability within established institutions.

15.5 Conclusions

The academic offerings in bioethics across Ibero-America show a continuous evolution and present a notable diversity that spans from short-term programs, such as courses and diplomas, to advanced levels, including master's and doctoral programs. This landscape reflects the increasing interest in bioethics training, addressing both immediate needs and long-term academic growth. The popularity of short-term programs is linked to their accessibility and responsiveness to educational demands, while the limited availability of advanced-level programs highlights challenges related to equity in access to higher education and resource availability.

Teaching modalities, particularly the flexibility of online formats, have made courses and diplomas the most common and relevant programs, especially in contexts where geographic or economic barriers impede face-to-face education. However, at advanced levels, such as master's and doctoral programs, in-person formats remain predominant, emphasizing the importance of direct interaction and collaborative learning in these academic settings.

The thematic focus also reveals significant trends: while clinical bioethics predominates in the bioethical landscape, thematic diversification toward other areas such as environmental and socio-legal ethics provides an opportunity to expand educational offerings to tackle global challenges like climate justice and sustainability, which are crucial in the current context. Additionally, socio-legal bioethics retains a relevant presence, underscoring its importance in promoting justice and human rights.

The professionalization of bioethics is another prominent trend, particularly in countries such as Argentina, Chile, and Mexico, where programs focus on practical application in clinical and hospital settings. In contrast, countries like Spain and Brazil demonstrate a stronger emphasis on academic training, highlighting the need to balance both perspectives to achieve a comprehensive education that integrates theoretical rigor with practical application. This highlights the importance of analyzing the factors influencing these trends more thoroughly. It is essential to understand the cultural, socioeconomic, and educational conditions that shape these orientations to create more balanced frameworks that combine theoretical rigor with practical application.

In this regard, analyzing how educational policies, labor market demands, and the academic traditions of each country contribute to these differences is essential. Additionally, identifying barriers that limit the adoption of hybrid models, such as a lack of resources, resistance to change, or inequalities in access to education, could provide tailored solutions to each region's particularities. A balanced approach between theoretical and practical training would not only enhance professional preparation but also promote regional collaboration and the establishment of shared standards in bioethics, thereby increasing its impact in key areas such as health, the environment, and social justice. Therefore, interdisciplinary research that incorporates pedagogical, social, and ethical perspectives is crucial for advancing toward a more inclusive and coherent bioethical education throughout the region. Regarding the relevance of the bioethics programs offered, moving toward more balanced frameworks that explore the integration of hybrid models, combining the accessibility and adaptability of digital education with academic depth and in-person collaborative networks, would be pertinent. Furthermore, it is vital to consider the contextual differences in Ibero-America and foster inter-institutional alliances that promote equity and sustainability in bioethics training in the region.

Based on the previous analysis, a series of proposals are presented for future research and the expansion of academic offerings:

1. Expansion in Environmental Topics: The low representation of this area in academic offerings presents an opportunity to develop programs that integrate environmental bioethics, with a particular focus on themes such as climate justice and ethics concerning the use of natural resources.
2. Hybrid Modality: Increasing the variety of hybrid programs could improve access to bioethics programs, particularly in areas where in-person education is difficult to maintain. This modality would blend the advantages of face-to-face interaction with the convenience of an online format.
3. Focus on Global Bioethics: Given the worldwide aspects of numerous bioethical dilemmas, integrating programs in English or with international content can enhance academic collaboration and draw students from various cultures.
4. Strengthening Interdisciplinarity: While the clinical focus is predominant, integrating perspectives from other disciplines, such as law, sociology, and environmental sciences, is essential for enhancing a comprehensive understanding of bioethics.

This study aims to act as a mechanism for the necessary continuity in the formative analysis of future strategies for bioethics education in Ibero-America, directing efforts toward a more inclusive and diverse academic offering that aligns with contemporary needs. The academic offerings in bioethics in Ibero-America are constantly evolving, addressing current challenges and anticipating future needs. Strengthening the inclusion, diversity, and quality of these programs is essential for consolidating their role as a key tool in promoting global and sustainable ethics.

References

1. Coromoto BS. La Bioética como Eje Transversal en el Pensum de Estudio de la Facultad de Odontología de la Universidad de Zulia, Venezuela. Acta Odontológica Venez. 2013:51.
2. Sierra Trujillo DC. Bioética y evolución del desarrollo moral en estudiantes de medicina. Análisis de caso en la Escuela de Medicina de la Universidad Pedagógica y Tecnológica de Colombia. [masthe's thesis]. Bogotá: Pontificia Universidad Javeriana; 2016.
3. Molina RN. La bioética: sus principios y propósitos, para un mundo tecnocientífico, multicultural y diverso. Rev Colomb Bioét. 2013;8:18–37.
4. Ferrari AG, Silva CMD, Siqueira JED. Ensino de bioética nas escolas de medicina da América Latina. Rev Bioét. 2018;26:228–34. https://doi.org/10.1590/1983-80422018262243.
5. Villegas-Múnera EM. Significados sobre bioética que surgen en un currículo de medicina. IATREIA. 2016:29. https://doi.org/10.17533/udea.iatreia.v29n3a08.
6. Oliveira ELD, Dantas AGDA, Britto Filho CO, Jacob MCM. Falhas de comunicação e epistemológicas limitam a abordagem da Bioética na formação e na prática em Saúde. Interface Comun Saúde Educ. 2021;25:e200679. https://doi.org/10.1590/interface.200679.
7. Cardona Ramírez CM. El desarrollo de la bioética en latinoamerica y el aporte desde el franciscanismo. El Ágora USB. 2010;10:243–8.
8. Abellán Salort JC, Maluf F. De la enseñanza de la bioética a la educación bioética. Reflexiones sobre los desafíos en los posgrados en Bioética. Rev Latinoam Bioét. 2014;14:52–65.
9. Cantú Martínez PC. Bioética y educación superior en México. Acta Bioethica. 2015;21:45–52. https://doi.org/10.4067/S1726-569X2015000100006.

10. Vidal S. La Educación en bioética en América Latina y el Caribe. UNESCO; 2012.
11. Susana V, Manchola-Castillo C, Garrafa V, Cárdenas Cepero YL, Ribot Guzmán E, Donda C, et al. Manual de educación en bioética: la agenda curricular en bioética: abriendo horizontes, volumen 1. Montevideo/Ciudad de México: UNESCO/Universidad Nacional Autónoma de México; 2021.
12. Pfeiffer ML, Manchola-Castillo C, Albuquerque A, Cruz E, Fabri dos Anjos M, Fonti D, et al. Manual de educación en bioética: fundamentar la bioética: conocimientos, valores y visiones desde América Latina y el Caribe, volumen 2. Montevideo/Ciudad de México: UNESCO/Universidad Nacional Autónoma de México; 2022.
13. Rueda-Castro L, Donoso-Flores I, Lagos-Bosman K, Briceño-Ribot C. Enseñanza de la bioética, generación e implementación de cursos en la Universidad Regional de O'Higgins (Chile). Apunt Bioét. 2023;6:29–46. https://doi.org/10.35383/apuntes.v6i1.916.
14. Fischer ML, Cunha TR, Moser AM, Diniz ALF. Metodologias inovadoras no ensino da bioética para o curso de Licenciatura em Ciências Biológicas. Rev EDaPECI. 2018;18:128–42. https://doi.org/10.29276/redapeci.2018.18.28052.128-142.
15. Gómez Urrutia V, Royo UP. Construyendo espacios de diálogo: una propuesta de enseñanza de la bioética. Acta Bioethica. 2015;21:9–18. https://doi.org/10.4067/S1726-569X2015000100002.
16. Rodríguez Escobar G, Sánchez Alfaro LA, Pinto Bustamante BJ. Ética Narrativa: un escenario para la deliberación bioética. Rev Colomb Bioét. 2016;11:92. https://doi.org/10.18270/rcb.v11i1.1815.
17. Torres Mendoza MV, Zepeda Gutiérrez DO. Revisión documental de las publicaciones en la enseñanza de la bioética en facultades de medicina de Latinoamérica. Rev Científica Estelí. 2024:26–41. https://doi.org/10.5377/esteli.v13i1.17705.
18. Sayago M, Amoretti R. Comités de bioética hospitalaria: importancia, funcionamiento y dificultades de implementación. Rev Bioét. 2022;29:832–43. https://doi.org/10.1590/1983-80422021294517.
19. Tealdi JC, Mainetti JA. Los comités hospitalarios de ética. Bol Oficina Sanit Panam. 1990;108:431–8.
20. Centro Anáhuac de Desarrollo Estratégico en Bioética (CADEBI). Observatorio Iberoamericano de Bioética (ORIBI) [Internet]. Available from: https://adam.anahuac.mx/ORIBI/#/inicio

Chapter 16
Medical Simulation in Clinical Bioethics Education

Ariana Cerón Apipilhuasco

In 2018, the Ministry of Health reported 133,000 medical students in Mexico—a number that grows yearly. This raises important questions for educational institutions: How can we ensure that our students acquire the necessary competencies to meet the growing demands of the medical field? What innovative tools can we provide to help them excel? And, above all, how can we guarantee that the education they receive is of high quality and equips them with the skills needed to become highly competent physicians prepared for future challenges?

Today, patient harm due to unsafe care represents a major and escalating challenge for global public health and is one of the leading causes of death and disability worldwide. Most of these harms are preventable. The importance of a safety culture has been extensively studied over the past decade and emphasized by the World Health Organization, the National Patient Safety Agency, and the Agency for Healthcare Research and Quality. Nonetheless, the concept of patient safety culture is relatively underreported in Latin American health literature and particularly in Mexico, where mortality rates are among the highest due to unsafe patient care.

Health systems are remarkably complex, integrating networks of professionals, multiple settings, and a wide range of technologies. Moreover, these systems must adapt iteratively to both external and local challenges while ensuring patient safety. The demands placed on them continue to grow alongside societal expectations. The economic costs of safety failures are high, and public trust in local health systems often diminishes when such incidents become known. Health professionals involved in serious safety events resulting in a patient's death or severe harm can also experience lasting psychological distress, along with deep feelings of guilt and self-criticism [1].

A. C. Apipilhuasco (✉)
Universidad Anahuac, Ciudad de México, México
e-mail: ariana.ceron@anahuac.mx

D. Cerdio-Domínguez, M. E. de los Ríos-Uriarte (eds.), *Current Bioethics*, New Paradigms in Healthcare, https://doi.org/10.1007/978-3-032-25294-4_16

Many causes negatively impact patient safety. Among them are insufficient training of healthcare staff, nonadherence to each institution's regulations, lack of interprofessional education, and unfamiliarity with communication tools—which are fundamental to a safety culture. Effective communication is indispensable for sharing information about patients' status, progress, and needs and for coordinating teams and tasks. High-performance teams require a closed-loop communication system so that instructions and specific roles remain clear and efficient, positively impacting patient care.

Strategies have been developed to help us identify our areas for improvement, providing opportunities for reflection, support, and the inclusion of everyone's perspectives. The goal is to change or refine processes to reduce patient mortality and improve the work environment [2].

Healthcare is a field in constant innovation and rapid evolution, aimed at enhancing services and overall quality of life. Therefore, we need highly trained and competent health professionals to meet medical demands and competencies. "Simulation" is a broad term describing the controlled reproduction of real processes for educational purposes through hands-on learning. Unlike traditional learning, Medical simulation allows clinical skill acquisition via deliberate practice.

Simulation tools offer an alternative to practicing on real patients, enabling learners to develop mental dexterity and responsiveness in patient care, with timely correction and enhanced competencies. Simulation-based education employs advanced technologies, including high-fidelity manikins and virtual reality, to recreate realistic clinical scenarios. These simulations allow students to engage in various medical situations—from routine procedures to complex emergencies without risking harm to real patients. This experiential learning improves technical competence and fosters critical thinking, teamwork, and essential communication skills for effective patient care.

Conducting an introduction at the start of a simulation or any event is potentially a critical mitigation phase that prevents challenging debriefing scenarios. Some limitations of simulation must be proactively acknowledged by facilitators during the initial briefing or at the opening of the debriefing and accepted by learners under the "fiction contract." Yet facilitators still face four distinct challenges:

- *Novice facilitators*: Each session offers an opportunity to train in the complex debriefing process and develop facilitation skills, especially communication. Early establishment of acceptance, trust, confidentiality, active listening, curiosity, and inclusive discussion with open-ended questions must be practiced continuously during reflection, when participants can express emotions and suggestions for improvement. Cognitive tools such as a brief "plus-delta" self-assessment after each practice, or recording oneself to enhance communication, help maintain focus. Reflection questions reveal learners' mindsets.
- *Structuring the debriefing*: The debriefing design dictates how much time is devoted to reflection and ensures that the session meets its objectives.
- *Expert participants*: A debriefing can become complicated if an expert learner challenges the session, which can frustrate inexperienced facilitators. Questions

must be carefully chosen and directed; otherwise, the debriefing can lose seriousness and focus.

- *Complex participants*: Some learners may be difficult to manage, especially if they have had negative experiences during the session, which adds another layer of challenge for debriefers [2].

Even with an excellent briefing and careful attention to participants' backgrounds, culture, and environment, facilitating the conversation can be complicated if a learner is unable or unwilling to engage in reflective dialogue. Such individuals may require a different approach, perhaps a more interactive format with a smaller group, since some feel exposed speaking in public. Maintaining psychological safety, closely tied to emotions evoked during simulation, is essential.

Psychological safety is now recognized as critically important because emotions play a significant role in enabling deep, valuable reflections. Without this environment, effective debriefing cannot occur. Thus, recognizing and validating emotions are a considerable challenge, especially in healthcare, where the activities themselves are emotionally charged. Yet it is precisely here that extra care is needed to foster learning.

Among simulation tools, the debriefing is the most impactful phase for analyzing each participant's learning framework. Medical simulation debriefing is vital for students to consolidate knowledge gained during simulations. It is a discussion between two or more people reviewing simulated events. Participants analyze their behaviors, thought processes, and emotions experienced during the simulation. Through proper debriefing, participants can improve aspects they identified during the exercise [3].

Debriefing fosters deep learning and increases the likelihood of transferring skills to the clinical environment. In adults, debriefing promotes continuous reflection on clinical actions and encourages improvement-oriented thinking among health professionals. High-quality debriefing requires active student participation to achieve profound learning and reflection, not only on clinical/theoretical decisions but also on ethical and empathetic choices [4]. A study by the Public University of Navarra concluded that debriefing helped students develop communication skills in clinical interviews, as well as strategies such as active listening, empathy, emotional engagement, offering alternatives to patients, and a more holistic approach to clinical practice [5, 6].

Another debriefing application is post-shift, post-rounds, or post-event sessions in healthcare, which encourage joint reflection, collaborative learning, and continuous adjustments within teams, thus improving both team processes and patient outcomes. Discussion topics should be achievable in the short- to medium-term and resolvable without excessive effort, so staff remain motivated and engaged; as a result, changes involve all personnel and become evident in practice [6].

Maintaining a culture of respectful communication centered on patient safety also positively affects the financial performance of healthcare institutions, underlining its crucial role in ongoing team development, system improvement, quality optimization, and sustainability, with the potential to reduce medical errors and omissions. Reflective debriefing after a critical event can also help decrease burnout and improve staff well-being and resilience.

Effective communication across teams after a pandemic such as COVID-19 is crucial for both team well-being and organizational efficiency, especially as the adoption of hybrid or remote work models accelerates. Clear, constant communication enables teams to adapt efficiently to these changes and promotes collaboration despite physical distance [7].

Emotional well-being and mental health must be preserved, given that stress and anxiety related to the pandemic affected the majority of healthcare workers. Open, frequent communication among teams provides a space to express concerns and receive support. Coordination and alignment during and after critical events, when organizations often revise strategies, procedures, and goals, depend on fluid communication to ensure that all team members remain aligned with new objectives and processes [8].

Strengthening the team through trust, training, and constant communication fosters mutual support, a culture essential for meeting future challenges [9].

However, debriefing can be complicated by cultural differences. The concept of reflective learning in debriefing is rooted primarily in Western cultures; all cultures have unique preferences, practices, and norms of teaching and learning. Facilitators must adopt culturally sensitive interpretations of simulation-based learning to ensure that students receive the maximum benefit from debriefing.

While debriefing focuses on enhancing clinical and communicative skills, it also serves as a vital space for ethical reflection. When bioethics intertwines with lessons from simulation, it offers students a moral compass to guide their future patient-care decisions, ensuring adherence to core medical principles. In this context, bioethics complements clinical decision-making by considering not only medical outcomes but also the values, principles, and responsibilities that underpin professional practice. Debriefing provides a safe environment where students can reflect on decisions affecting patients and team dynamics, acquiring not only effective communication skills but also empathy and informed decision-making crucial for addressing ethical challenges in clinical practice. Principles such as patient autonomy, beneficence, justice, and non-maleficence are integrated into decision-making. In summary, embedding bioethics into medical simulation debriefing enriches technical learning and produces health professionals equipped to face complex ethical challenges in practice [10].

Traditional teaching environments centered on theoretical knowledge often neglect the deep development of ethical competencies needed to tackle dilemmas beyond the clinical aspect. Although medical simulation has proven effective for reinforcing clinical skills, its true potential lies in debriefing. This field remains open for exploration and research to achieve full integration of ethical reflection into every simulation experienced by students. The challenge is how to optimize debriefing to foster ethical reflection, thereby preparing students not only for clinical practice but also for ethical responsibility [11].

Research shows that, regardless of how frequently students participate in simulations, they consistently value debriefing as a key educational tool. Palés Argullós et al. [12] highlight that debriefing provides an opportunity for reflective learning, a feature appreciated regardless of prior simulation experience. A significant relationship has been found between simulation participation and perceived debriefing effectiveness, suggesting that students with more simulation experience perceive debriefing as more effective. This supports the idea that repeated simulations improve reflective capacity and heighten the perceived value of debriefing for ethical decision-making.

Similarly, studies [13] have demonstrated that debriefing offers an ideal platform for ethical reflection and complex decision-making in clinical contexts. Leiphrakpam et al. [11] also report that medical simulation enhances students' ability to confront ethical dilemmas in practice.

Interestingly, no significant relationship has been found between perceived debriefing effectiveness and comfort in discussing ethical issues during debriefing. This may be because technical effectiveness does not necessarily influence students' comfort in addressing ethics, which depends more on the trust-building environment created by the facilitator than on the debriefing structure itself [14].

Debriefing fosters ethical reflection by providing a safe space for discussing moral dilemmas encountered in simulations. Results suggest that, even if not all students face ethical dilemmas during simulation, those who do find debriefing effective for ethical reflection—aligning with prior studies [11].

It is essential to recognize that the focal point is the student, whose response will depend on the activity type, their background, and the evaluation's purpose. Debriefing is a beneficial process to engage participants after any simulation or learning activity, regardless of modality. Importantly, facilitators need not be simulation experts; the goal can simply be to guide learners through direct or indirect reflection. "Direct vs. indirect guidance" refers to whether someone actively facilitates the process or participants rely on a provided framework for self-debriefing, feedback, or peer-to-peer feedback. Depending on the chosen debriefing approach, this process can reveal mental models underlying behaviors and cognitive reasoning, which can then be corrected or adjusted for future performance improvement.

Throughout the simulation, we seek not only emotional well-being but also psychological safety to foster genuine reflection and motivate change for the sake of patient welfare and a better work environment.

16.1 Psychological Safety in Simulation and Debriefing

Informative sessions should promote reflection and help learners make sense of events. Threats to psychological safety can undermine reflective learning conversations and inhibit the transfer of key lessons from simulated cases to real-world patient care. Effective debriefings require high levels of psychological safety: the perception that it is safe to take interpersonal risks without fear of embarrassment, rejection, or punishment for expressing ideas, admitting lack of knowledge, or asking questions [15].

Highly effective debriefings help learners make sense of events and, through reflection, promote the transfer of learning from simulations to general patient care contexts [16].

It is common for healthcare personnel to make mistakes when performing a new procedure or one they haven't practiced recently. Facilitators should allow reflection on these errors. However, in reality, organizational culture often treats errors as events to be avoided and punished, without providing effective feedback to sustain improvement. Successes may be rewarded, but without follow-up, the process cannot continue improving—worsened by a lack of clarity about proper procedures, which fosters insecurity, fear, and frustration [17].

Research on organizational behavior shows that psychological safety depends on factors at three levels: individual, team, and organization. For example:

- *At the individual level, the traits are* proactive personality (willingness to engage in proactive behavior despite external forces), emotional stability (tendency to feel calm, relaxed, and steady), and learning orientation (focus on skill development rather than on demonstrating high performance).
- *Team level*: inclusive leadership (leaders' words and actions that invite and value contributions), work design features (role clarity, interdependence, autonomy), peer support, and mutual trust and respect.

Psychological safety is a subjective, complex perception that cannot be "switched on" or declared simply by labeling an environment as "safe." It is dynamic and fragile, especially in interprofessional and interdisciplinary contexts where not all team members may experience the same level of safety at any given moment. For instance, physicians in a team briefing may feel more psychologically safe with one another than do nurses, or vice versa [18].

Psychological safety supports creativity, engagement, performance, information sharing, expression, and learning. It is therefore a necessary condition for effective team functioning, as members must trust that one another's work is supported, especially during critical actions. Yet organizational cultures, both in healthcare and beyond, often implicitly view errors as "crimes" deserving punishment or avoidance.

In some healthcare settings and specialties, behaviors such as mistreatment, belittlement, and harsh criticism persist under a "culture of abuse" in nursing education. Morbidity and mortality conferences can be seen as venues of "shame and blame" rather than learning. Leaders may reward staff for not making errors or for quickly correcting them, but not for discussing them. Consequently, when errors occur or when staff seek help or feedback, they understandably feel fear, anxiety, and shame. They may withdraw, avoid asking for help, and conceal mistakes. Ambiguous stimuli are more likely to be seen as threats when learners are already anxious, because emotions frame how people process information. These findings explain why innocent actions by debriefers can inadvertently trigger defensive or withdrawn responses. Learner engagement in debriefing may hinge on how educators address anxiety, frustration, fear, reluctance to participate, or more dramatic breaches of psychological safety. Learners must feel safe to be vulnerable and take interpersonal risks such as admitting ignorance or mistakes. Without this honesty, educators may miss important reasons behind performance gaps. Psychological safety encourages learners to "try and fail at the edge of their experience, where knowledge and skills may or may not suffice to prevent errors."

The tension between fostering psychological safety for learning conversations and the broader organizational norms that suppress it poses a challenge for many debriefers. Rather than simply reinforcing existing organizational practices, debriefers may feel compelled to create an "island of safety" during debriefings. While this complicates their work, it can motivate educators to identify strategies for enhancing psychological safety in briefing sessions. This challenge is even greater in hospital settings, where stress levels are high, but simulation-based training can help improve educational environments [19].

References

1. Díaz Navarro AR. Evaluación of TALK training for professional clinical debriefing in Latin America. Arch Med Res. 2024:55–69.
2. Eppich W, Cheng A. Promoting excellence and reflective learning in simulation PEARLS: development and rationale for a blended approach to healthcare simulation debriefing. Simul Healthc. 2015:106–15.
3. Institute for Medical Simulation. Manual de trabajo EDSS completo. Harvard Medical School; 2016. https://harvardmedsim.org/wp-content/uploads/2017/06/Manual-de-trabajo-EDSS-completo-2016-agosto.pdf
4. Loría J. Debriefing en simulación clínica. ResearchGate; 2021. https://www.researchgate.net/profile/Jorge-Loria-2/publication/353687946_Debriefing_en_simulacion_clinica/links/610a92510c2bfa282a20c955/Debriefing-en-simulacion-clinica.pdf
5. Arenas-Monreal L, Bárcenas-López R, Gutiérrez-Pérez L. La ética en la simulación clínica: revisión sistemática. Contextos Educativos Revista de Educación. 2020;26:199–217. https://doi.org/10.18172/con.5126.
6. Motola I, Devine LA. Simulation in healthcare education: a best evidence practical guide. Medical Teacher; 2013. p. 511–30.
7. He M, Harris L. The impact of COVID -19 pandemic on corporate social responsibility and marketing philosophy. J Bus Res. 2020:176–82.
8. Dirani KA. Leadership competencies and the essential role of human resource development in times of crisis: a response to COVID 19. Hum Resour Dev Int. 2020:380–94.
9. Seeger MW, Pechta LE. A conceptual model for evaluating emergency risk communication in public health crises. Health Secur. 2020:246–55.
10. Rodríguez Weber FL, Ortega Cerda JJ, Ramírez Arias JL. Bioética y su aprendizaje en el equipo de salud. Acta médica Grupo Ángeles. 2018;16(1):5–6. http://www.scielo.org.mx/scielo.php?script=sci_arttext&pid=S1870-72032018000100005&lng=es&tlng=es
11. Leiphrakpam PD, Armijo PR, Are C. Incorporation of simulation in graduate medical education: historical perspectives, current status, and future directions. Simul Healthc. 2024;20(1):123–9. https://doi.org/10.1177/23821205241257329.
12. Palés Argullós JL, Gomar Sancho C. El uso de las simulaciones en educación médica. Teoría de la Educación. Educación y Cultura en la Sociedad de la. Información. 2010;11(2):147–69. https://www.redalyc.org/pdf/2010/201014893008.pdf
13. Emmerich N, Gormley G, McCullough M. Ethics of healthcare simulation. In: Healthcare simulation education; 2017. p. 119–26. https://doi.org/10.1002/9781119061656.ch16.

14. Maestr n-based medical teaching e JM, Rudolph JW. Teorías y estilos de debriefing: El método con buen juicio como herramienta de evaluación formativa en salud. Rev Esp Cardiol. 2014;67(7):567–70. https://doi.org/10.1016/j.recesp.2014.05.018.
15. Kolbe M, Eppich W. Gestión de la seguridad psicologica en las sesiones informativas: un acto de equilibrio dinámico. Simul Technol Enhanc Learn. 2020:164–71.
16. Cheng A, Eppich W. Debriefing para la simulación mejorada por la tecnología: una revisión sistemática y metaanálisis. Med Educ. 2014:657–66.
17. Schein EH. ¿Como pueden las organizaciones aprender más rápido? El reto de entrar en la sala verde. Sloan Manag Rev. 1993:85–92.
18. Edmondson AL. Psychological safety: the history, renaissance, and future of an interpersonal construct. Annu Rev Organ Psych Organ Behav. 2014:23–43.
19. Rudolph JW, Raemer DB. Establishing a safe container for learning in simulation: the role of the presimulation briefing. Simul Healthc; 2014. p. 339–49.

Recommended Bibliography

Agency NP. How to guide to the five steps to safety surgery, five steps to safer surgery. National Patient Safety Agency; 2010. p. 12–20.

Al-Elq A. Simulation-based medical teaching and learning. J Fam Community Med. 2010;17(1):35. https://doi.org/10.4103/1319-1683.68787.

Antoni M, Zentner J. Las cuatro emociones básicas. In: Antoni M, editor. Las cuatro emociones básicas. España: Editorial; 2020. p. 30–56.

Arriaga AF, Sweeney RE. Failure to debrief after critical events in anesthesia is associated with failures in communication during the event. Anesthesiology. 2019:1039–48.

Ayala JL, Romero LE. La simulación clínica como estrategia de enseñanza-aprendizaje en ciencias de la salud. Metro Ciencia. 2019:32–8.

Cerón A, Rodríguez-Cruz L. Introducción a la simulación clínica. Rev Educ Investig Emerg. 2020:140–4.

Choudhry NK, Fletcher RH. Systematic review: the relationship between clinical experience and quality of health care. Ann Intern Med. 2006:260–73.

Chung A, S. B. Exploring new methods in simulation science. J Simul Stud. 2013:123–45.

Collings AS. Checklists, surgical timeout, briefing, and debriefing: safety in the operating room. In: The SAGES Manual of Quality, Outcomes, and Patient Safety. Springer; 2022. p. 20–8.

Edmondson AC. Psychological safety: the history, renaissance, and future of an interpersonal construct. Rev Org Anual Psyc Organ Behav. 2014:1–23.

Foulk T, Woolum A. Catching rudeness is like catching a cold: the contagion effects of low- intensity negative behaviors. J Appl Psychol. 2016:23–45.

Gómez P, M R. Salud emocional y binestar en profesionales de la salud. Médica Internacional. 2019:20–9.

Jiya T. Ethical reflections of human brain research and smart information systems. ResearchGate; 2019. https://www.researchgate.net/publication/333781242_Ethical_Reflections_of_Human_Brain_Research_and_Smart_Information_Systems

Levett Jones LS. A systematic review of the effectiveness of simulation debriefing in health professional education. Nurse Educ Today. 2014:58–63.

López Gómez MI, Tarasco Michel M. Relación médico-paciente. In: J. J. García (Dir.), Enciclopedia de Bioética; 2020.

López S, López LR. La simulación clínica como herramienta de aprendizaje. CIR MAY AMB. 2012:25–9.

Kohn LT. Errar es ser humano construyendo un sistema de salud más seguro. National Academy Press; 2000. p. 1–25.

Madan AK, Caruso BA, Lopes JE, Gracely EJ. Comparison of simulated patient and didactic methods of teaching HIV risk assessment to medical residents. Am J Prev Med. 1998;15(2):103–8. https://doi.org/10.1016/s0749-3797(98)00026-9.

O'Donell JMF. Debriefing: a critical component in simulation. Clin Simul Nurs. 2009:55–7.

Pablo Fernández-Berrocal DR. La Inteligencia emocional en la Educación. In: Fernández-Berrocal DRP, editor. La Inteligencia emocional en la Educación. España: PsicoSalud; 2021. p. 421–36.

Raemer DB, Anderson M. Research regarding debriefing as part of the learning process. Simul Healthc. 2011:52–7.

Sanchez L, P. A. Estrategias de manejo emocional en el personal sanitario. Revista de psicología y salud ocupacional. 2020:45–58.

Secretaria de Salud. 239. se incrementa en 43 mil el número de estudiantes de medicina en los últimos 8 años [Internet]. Gobierno de México; 2018 [cited 2024 Jul]. Available from: https://www.gob.mx/salud/prensa/239-se-incrementa-en-43-mil-el-numero-de-estudiantes-de-medicina-en-los-ultimos-8-anos

So HY, Wong GKC. Simulation in medical education. J R Coll Physicians Edinb. 2019:52–7.

GPSR Compliance

The European Union's (EU) General Product Safety Regulation (GPSR) is a set of rules that requires consumer products to be safe and our obligations to ensure this.

If you have any concerns about our products, you can contact us on ProductSafety@springernature.com

In case Publisher is established outside the EU, the EU authorized representative is:

Springer Nature Customer Service Center GmbH
Europaplatz 3
69115 Heidelberg, Germany

Batch number: 10726060

Printed by Printforce, the Netherlands